The World Connection

Timothy Orr Knight is a member of the "younger generation" who started playing around with computers three years ago. He has "fallen in love" with them, however, and is now taking them very seriously. This has developed into a strong interest in "modems," the devices that allow one microcomputer to talk to another computer.

In addition to making plans to attend college, Tim has been extremely busy programming computer games, such as *EVADE, SPACE CHASE,* and *SPOX,* for various software companies while also writing reviews, articles, and programs for such national magazines as *InfoWorld* and *Softside*. He plans on writing two more books before starting to college.

Timothy Orr Knight is a busy 16-year-old who enjoys being with friends and being involved in all areas of school activities, but most of all . . . , he enjoys **working.** Thus, it is no surprise that Tim plans on forming his own software company in the spring of 1983, producing software for the IBM-PC and the TRS-80.

The World Connection

by
Timothy Orr Knight

Howard W. Sams & Co., Inc.
4300 WEST 62ND ST. INDIANAPOLIS, INDIANA 46268 USA

Copyright © 1983 by Timothy Orr Knight

FIRST EDITION
FIRST PRINTING—1983

All rights reserved. No part of this book shall be reproduced, stored in a retrieval system, or transmitted by any means, electronic, mechanical, photocopying, recording, or otherwise, without written permission from the publisher. No patent liability is assumed with respect to the use of the information contained herein. While every precaution has been taken in the preparation of this book, the publisher assumes no responsibility for errors or omissions. Neither is any liability assumed for damages resulting from the use of the information contained herein.

International Standard Book Number: 0-672-22042-3
Library of Congress Catalog Card Number: 82-61969

Edited by: *Frank N. Speights*

Printed in the United States of America.

Preface

This book is mainly about *modems,* those devices that allow computers to communicate with other computers. I decided to write this book because I felt that the old saying of "Find a need and fill it" was true.

I began playing around with computers about three years ago, when I was 13 years old and my interest developed into a strong desire to better understand computers and their peripherals, especially modems, because I felt that computer communications would be very important in the coming years. To me, the question "How is the area of computer communications going to affect my life?" was very important.

However, I could not find a book that discussed computer communications—what I call "the *new* world connection." Therefore, I decided to write this book to try and show people what I believed was going to occur, how they could prepare for it, and what they might want to buy in the line of computer hardware and software to be ready for this new world.

In the earlier parts of *The World Connection,* I try to explain how modems operate and I try to answer the question of how computer communications is going to affect our lives. Time-sharing computer systems, like *The Source* and the *CompuServe Information Service,* will be extensively covered. Bulletin Board Services, or the little systems, will also be examined, along with a discussion of the people who own and run the bulletin board services—the system operator, or *SYSOP.* In addition to modems, other hardware devices, such as smart and dumb terminals, will be examined. Electronic mail and other technical advances will be discussed, as will another new problem—the computer "pirate."

Finally, I review some of the available software and present some of my thoughts on the effect of technology and what I think will be some of the things to come. I end the book with a glossary of terms, a list of the bulletin board service numbers, and a list of the addresses of some hardware and software suppliers.

I hope my efforts will ease your way when you finally decide to take the plunge and hook your microcomputer to a modem. I hope your first step into *The World Connection* via the time-sharing computer systems is a pleasure because this book has removed all the pitfalls from your path.

TIMOTHY ORR KNIGHT

This book is dedicated to my mother and father, who are easier to talk with than any computer in the world.

Acknowledgements

I would like to thank the many companies that supplied photographs, software, information, and other assistance to me in the creation of this book. The products of these companies and all of the others mentioned in this book are trademarked by their owners. I give my sincere thanks to:

    Alternate Source (Modem 80)
    Instant Software, Inc. (Super Terminal)
    Lance Miklus, Inc. (ST-80 series)
    Lear Siegler, Inc. (Terminals)
    Lindbergh Systems (OMNITERM Software)
    Novation, Inc. (maker of the CAT modems)
    Small Business Systems Group, Inc. (ST-80 and FORUM-80 software supplier)
    TYMNET, Inc. (a common carrier)

I want, also, to express my appreciation to Morrisa Sherman for the many fine sketches that I have used throughout this book.

Contents

CHAPTER 1

Computer Communications 11
   What This Book Is About—Big Guys and Little Guys—Rational Reviews—Be a SYSOP—Some Future Possibilities

CHAPTER 2

The World Connection Exposed 19
   The Communications Addiction—A Little Bit Technical—Clarifying Terms—Ma Bell and Friends — Paying the Price — The Business View — A Big Step Forward

CHAPTER 3

The Big Guys 30
   Defining the Networks—May the Source Be With You—Joining the Source—An Imaginary Session—A Look at CompuServe—The CompuServe Subject Index—Extras on CompuServe—Citizen's Band Radio—Life, Love, and Trivia—The Multi-Player Host—Summary—Final Words on the Big Guys

CHAPTER 4

The Little Guys 56
   What Is a BBS?—Calling the BBS—The Commands —Advantages of the BBS — Using a BBS Wisely— Future of the Bulletin Board Service

CHAPTER 5

Naughty, Naughty—An Expose 65
  Examples of Computer Crime—The Home Computer Criminal—Yo Ho Ho—How It's Done—What Is Being Done to Stop Piracy—Phone Phreaking—Where Will the Naughtiness End?

CHAPTER 6

A Multitude of Modems 75
  Terms To Look For—The LYNX®—Hayes Smartmodem®—Hayes Smartmodem 1200®—Hayes Micromodem II®—Hayes Terminal Program—Telephone Interface II—Direct-Connect Modems I and II—Modems From Novation, Inc.—The AUTO-CAT™ Modem—Apple-Cat II™—Your Buying Decision—Terminals—The Infone™

CHAPTER 7

Be a SYSOP 98
  Getting Started—Choosing a BBS for Yourself—Why Be a SYSOP?—The Drawbacks and Headaches—To Be or Not To Be?

CHAPTER 8

The Soft Side 108
  Reviewing the Terms—The Software—Things To Consider Before You Buy

CHAPTER 9

The World Connected 117
  Electronic Mail—The Software Industry—New Technology—Technology's Effect—Democracy Threatened—People and Computers—Final Thoughts on Things To Come

APPENDIX A

Glossary of Terms 127

APPENDIX B

Bulletin Board Service Numbers 131
  BBSs—Meanings of Names and Abbreviations

APPENDIX C

Addresses of Hardware/Software Suppliers 136

Index 139

Chapter 1
Computer Communications

Welcome to *The World Connection*. In this book, I hope to let you see something coming that is so revolutionary and so expanding that it is going to change the world. Sound pretty dramatic? It is, and I hope that by the time you finish this book, you will realize the full impact of what I have told you.

I wanted to write this book because I felt that the old saying of "Find a need and fill it" was a true one. I knew that computer communications would be very important in the coming years, but I simply could not find a book on this new *world connection*. Therefore, I decided to write this book to show people what was coming up, how they could prepare for it, and what they might want to buy in the line of software and hardware in order to be ready for it.

What This Book Is About

Some of my words might be a little confusing if you are not familiar with computer communications. Don't let this alarm you. One of the purposes of this book is to make words like *modem* and *computer communications* words that you can know and use. For now, just understand that these terms only refer to the connection of one computer to another. Not necessarily physically, but in a way which connects a *computer* to a *phone line* and, then, to another *computer*. You might want to think of the two computers as just two "people" talking with one another (Fig. 1-1). This will help bring something as complex as modems and computers down to the practical and graspable concept of a friendly conversation.

My main intention in this book is to make reading it a learning experience and, also, make it something that you will want to keep and use as a reference. In addition to this

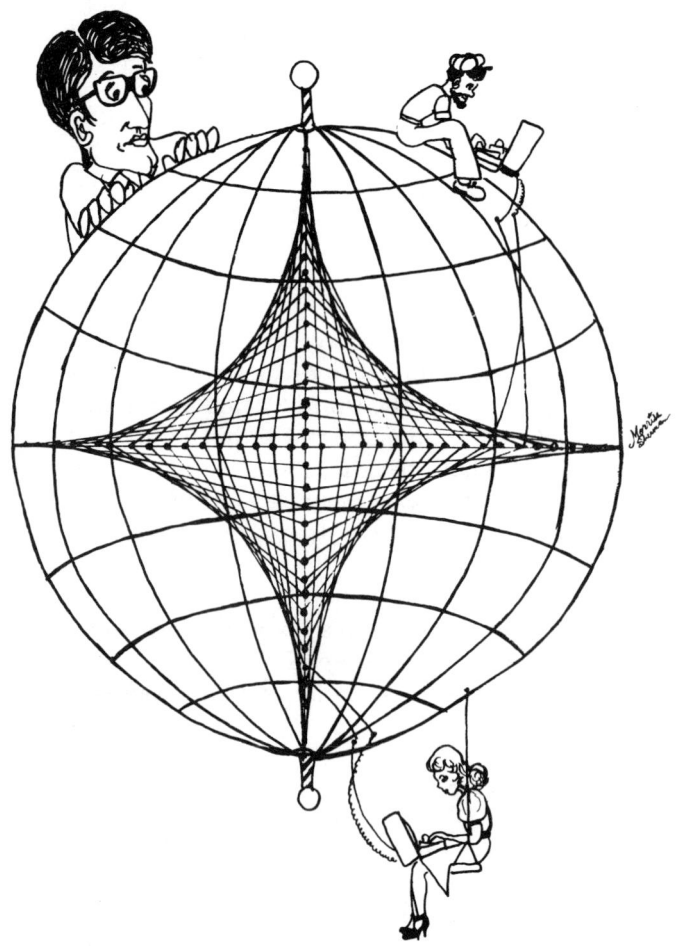

Fig. 1-1. Connecting the people of the world through computer communications.

"dual personality" for the book, which should make it much more useful to you, I have also tried to make sure that all levels of computerists can benefit from this book. If you don't know what a modem is or what a computer can do, you should still be able to understand and benefit from this book. On the other hand, if you have worked with computers for years and consider yourself a "top banana" in the field of computer communications, this book should still offer some new insights for you.

In the earlier parts of the book, I'm going to answer questions like "What is a modem?" and "How is the area of computer communications going to affect my own life?" However, an explanation of terms is very important to understanding an idea and so, for convenience sake, I have included a glossary in the appendices of the book for reference purposes.

Big Guys and Little Guys

You may have heard about some of the giant *time-share computer systems* which thousands of people call up every day. This will be the subject in the chapter entitled "The Big Guys." Huge computer systems like the *CompuServe Information Service* (Fig. 1-2) and *The Source* (Fig. 1-3) will be extensively covered. These constantly growing network systems, with tens of thousands of users, are popular, useful, and a tremendous amount of fun to use.

Fig. 1-2. CompuServe (Courtesy CompuServe Information Service).

In contrast to "the big guys," I'm also going to show a full-scale view of "the little guys" by covering the microcosm of what is commonly called *BBSs*. BBS stands for "Bulletin Board Service," and it is just that—a computer bulletin board which users may call, one at a time, and leave messages (bulletins) for other users. Fig. 1-4 shows the "menu" of a BBS. These BBSs are expanding both in number and in complexity and are beginning to serve as much more than just message centers. Many are bringing in purchasing sections from which users can order items "on-line." Even more of them are giving the ability to transmit programs to

Fig. 1-3. The Source (Courtesy Source Telecomputing Corp.).

Fig. 1-4. Using a bulletin board service.

and from the computer the BBS is run on. All of these things, and many others, will be covered in "The Little Guys" chapter, which explains how important the *smaller* computer connections will be.

14

Rational Reviews

One of the main reference parts of this book, which I think will help anyone who is considering getting into computer communications, is the hardware review section. Hardware, in case you don't know, is the electronic part of the computer system. It may be the computer itself, or a printer, a modem, or something else, but it is the *real working part* of any computer system. The hardware that I will review will be mostly *modems* because I believe they are the most important units in a system (Fig. 1-5). Modems are those devices that allow computers to communicate with each other; therefore, they are the key to the "world connection."

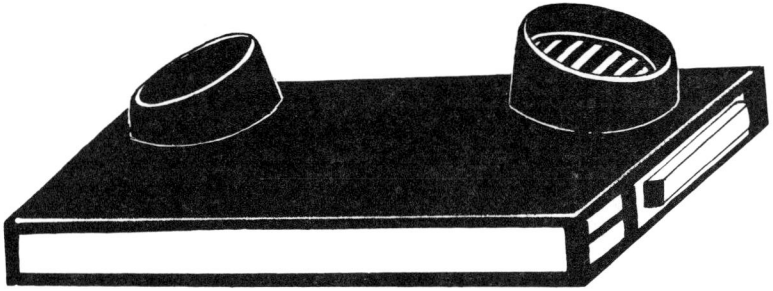

Fig. 1-5. A typical acoustic modem.

There are many modems on the market, so I have used not only my opinions but the opinions of *many* persons, in order to give a critical evaluation and a listing of the features of each type modem. Thus, you can decide for yourself which modem might be best for your own needs.

On the other side of the coin, I'm also going to review the pieces of software (the actual programs) that are involved with computer communications, and which are commonly called *smart terminal programs.* Any "smart buyer" knows that he or she must examine carefully what is available and, then, pick the item best suited for himself or herself. There is a tremendous amount of smart terminal software on the market these days and, again, I have tested much of it and have interviewed many others who have different types of software. Because of this, I can provide a strong critique for various software, which should help *anyone* who is trying to decide which package to buy.

There is one point I would like to make about both the hardware and the software reviews. These are not meant to be advertisements or biased praisings of *my* favorite types of hardware or software. I have collected the ideas and opinions of many people so that you can have a number of objective reviews at your disposal. The sole purpose of these reviews is to help potential buyers select the hardware and software that would be appropriate for them without having to use the "trial and error" method—a method that wastes quite a sum of money.

Be a SYSOP

Yet another aspect of the *world connection* which I shall cover is that of becoming a *SYSOP*. What is a SYSOP? Well, it stands for "SYStem OPerator." This is the person who owns and runs a bulletin board service (BBS).

The point here is that, perhaps, you might benefit from owning a BBS and being the SYSOP of it. It may seem a little far out at first thought, but it might just hold some strong benefits for you. Have you possibly been thinking about owning a software distribution center? A BBS is a fine way of reaching the *exclusive* computer public. Maybe you would just enjoy meeting more people with the same interests that you have. Again, being the SYSOP of a BBS is an excellent way of becoming known, since there are thousands upon thousands of computer users who call up a BBS every day of the year.

In case you're not so sure about whether you'd like to be a SYSOP or not, but you do have a modem and you would like to call up some BBS for a try, I have included a list of the current BBSs in the country (and one from *abroad*) in the appendices. There are currently well over 300 systems in the United States, and that number, I feel sure, will continue to grow.

In case that you *do* feel that a BBS would be something special for you, I have also included reviews of some of the major BBS software. This will help you to decide which BBS you would like most, and this is very important, since BBS programs vary greatly in quality, but not in price. They are all fairly expensive programs, so it would probably be best to look at these critiques and evaluations carefully before you go ahead and buy one. These reviews might even help those who just want to call up some good BBS and want to make sure they're calling one that they would enjoy.

Some Future Possibilities

On a slightly more "philosophical" level, I have included quite a bit in this book on what the *possibilities* of a "world connection" might be. I call it the "world connection" because I feel that is the most appropriate term for the *whole sphere* of computer communications. Also, whenever my friends ask me what good is my modem, I tell them that anyone who has one can "have the world at his fingertips," which can sometimes be pretty close to the truth (Fig. 1-6).

Fig. 1-6. The whole world at your fingertips.

I often find it interesting just to ponder what the growth of computer communications will do to the world that we know today. Looking at the little box we call a modem, one might not think of it as being so revolutionary. But, think of what might happen, though, through its use. I won't present any specifics of my own right now, but a couple of ideas that you might want to mull over in your mind are the areas of mail (electronic mail versus our present-day postal service) and how relationships between people just might change because of this new communications medium—the computer and its modem. I find possibilities like this fascinating and I will cover them extensively later in this book.

So let's begin our adventure. It may seem like there is a lot to cover with all the reviews, previews, and other information, but think of it as a preface of things to come, and as something that you as an *informed person* should be aware of. The whole idea of "world communications" is very thrilling and it may cause some exciting changes and improvements in your life and mine. We now will begin our journey into *The World Connection* . . . .

Chapter 2
The World Connection Exposed

"In *my* day, we didn't have television or radio, and I had to walk fifty miles to school in a blinding snowstorm without any shoes on . . . ."

If you are one of the millions of people who has heard some such story of suffering from your elders, and who are always being told about how *good* we all have it today compared to them, take heart—I have endured the same thing. Still, there is some merit to all of those tales of pain and peril in that many things *have* improved over the years, especially in the field of technology.

The Communications Addiction

For example, assume that for one month you had to live without your telephone. We know that humans lived for eons without telephones, so couldn't you do without one for just one month? However, you might wonder, "What would I do in an emergency? How could I call for help?", or perhaps you might ask, "What if I wanted to talk with one of my friends? I would have to drive miles just to do that." It seems that we have all grown dependent on the tools of communication, such as the telephone, and trying to live without them would be difficult at times.

Then, again, many people get most of their entertainment solely from television. Not all of them are "tv addicts," but they do watch their favorite shows and newscasts as often as they can. What would they do without their television set? What would they do with the time they used to spend watching tv? In the days before television, people read books, played parlor games, and used their inner resources, but once a person begins to rely on television as a source of entertainment, things like playing checkers may begin to seem very boring. Once again, here is a communication medium that many people have become to rely upon heavily and would find it hard to live without. The reason I am discussing the communication tools that we use so much today is to bring up the point that modems may very well become just as integral a part of our lives as our telephone or television is today. This may seem a bit far-fetched to the non-computer user, but I know from personal experience that this can happen. I had to be without my modem for about three months, and I found that it was very difficult for me at times, since a lot of my mail and entertainment came through the modem. This is good, since I discovered that modems are useful and I proved to myself that my modem was truly helping me. However, it is also bad because I have grown very dependent on a piece of computer equipment that I had done without for most of my life.

It could very well happen that in about ten years or so, people might be spending a little less time watching television and talking on the telephone and much more time using their computer to communicate with other computers and other people. This could be a very positive step if the main sources of information continue in the direction that they are headed now—education and entertainment. However, if most of these information bases rely on "fluff" to attract major audiences (like some television networks have done), it may not have a positive effect at all. These things remain to be seen.

A Little Bit Technical

This chapter is the only one which will be in any way technical, thought it won't be difficult at all to understand. I will be defining those words which may be confusing you (like *parity, bit, serial,* and so on), in addition to explaining both what the phone system is offering in terms of data communications and how you can use different types of phone systems to save yourself some money. Finally, I want to show how interested businesses are in this new communications medium. Therefore, this chapter isn't going to have anything to confuse you in it. In fact, it should clear up many of the things that are confusing you now.

First, I would like to reemphasize that the phone service is one of the most important parts of data communications (since it makes it all possible), so telephone systems will be mentioned frequently in this book.

The history of telephones is about a century old. Telephones have improved drastically over these past one hundred years, and the United States certainly has the best system in the entire world (even if you couldn't reach your mother by long distance during Christmas-time). Telephone communications have been made much more efficient and practical through such things as the long-distance *phone switching* system. If this computerized system did not exist, the Bell System would have to use all of its employees as operators.

Newer technological advances such as communications satellites, light-wave cables, and improved telephone transmissions are going to help the emerging data-communication systems in their efforts to grow and improve. At the present time, 750-million telephone calls are made every day. This number will increase as data transmissions become more widespread and more "domesticated."

Of course, the history of data communications is much less substantial than that of the telephone. We are near the beginning of that history, and our use of data communications may just be an integral part of the "history data banks" (not history books) of the future. That is one of the things that makes this emerging *world connection* so exciting.

Clarifying Terms

There are a lot of words that may have you a little confused. These terms are certainly nothing to be afraid of and the important ones will be clarified right now so that you can understand what is discussed from this point on.

Modem

A modem is the device that allows computers to communicate with one another (Fig. 2-1). Modems usually cost around $250 when used for a personal computer, and their communication medium is the telephone and the telephone lines. Modem is a short way of saying "MOdulator/DEModulator," which is what the modem is. It is a device that changes the language of the computer into electronic impulses which can be transmitted and which can also be received and translated back into the computer's language.

Fig. 2-1. An acoustic modem (Courtesy Novation, Inc.).

Acoustic

This word refers to "sound." Thus, an acoustic modem is the type that actually holds the handset of a telephone. To use this modem, you must dial a number, wait for a long high-pitched tone, and, then, push the handset into two rubber cups which act to shield the telephone from most outside noise. The computer can then talk (communicate) directly through the phone, just as people do. The advantage to this method is that acoustic modems are almost always cheaper than other types of modems. A disadvantage is that an outside noise, such as the slamming of a door, can cause an inaccurate data transmission. This can be very serious during an important transmission, such as a stock report or a business sale.

Direct Connect

This refers to a more expensive and reliable modem, or the direct-connect modem. This type of modem is connected with the telephone wire going directly to the modem and another wire going out of the modem to the telephone. The telephone may be used normally except when the modem is connected onto a computer system. Obviously, the advantage to a direct-connect modem is that any outside noise is almost nonexistent.

Bit

A *bit* is the smallest amount of computer data. "Bit" is short for "Binary digIT." Data are composed of bits that are made up (symbolically, if not electronically) of *ones* and *zeroes*. Think of a bit as being like a light bulb—it can be either on (1) or off (0). However, a bit inside a computer is not a light bulb but is, in fact, an extremely tiny switch in itself. The information is arranged in a way which would "look" like this if shown: 01010111 10110001 10100101. Each digit is a bit, and each set of eight bits composes a *byte,* the basic memory unit of a computer. Each byte usually accounts for a number, a symbol, or a letter of the alphabet, so the word "TIM" would take up three bytes of information, or 24 bits. That is a lot of light bulbs . . . .

Baud or bps

Baud is like a speed indicator. In a car, you can go a certain speed, such as 55 miles per hour. With modems, the "speed" of data transmission is measured in how many bits can be transmitted every *second.* Baud is the same as *bps,* which means bits per second. The standard rate of a personal computer modem is 300 baud, which is like 300 "ons" and "offs" every second. This baud rate suits most people quite well though, for some businesses, baud rates much faster (like 19,200 baud) are essential for a maximum data transmission in the minimum amount of time.

Full and Half Duplex

What should be understood here, about duplex operation, is that the keyboard (the input) and the video display (the output) are actually two separate devices (Fig. 2-2). As far as computer communications goes, you may not always see on the video screen everything you type on the terminal. That is exactly what full and half duplex operation is for. In some communications, each end of the system can transmit and receive "messages" simultaneously (full duplex). However, in other systems, technical requirements may permit half duplex, or one-direction, communications between stations. The "communications" can be in either direction, but not simultaneously.

Fig. 2-2. Computers can be operated either full or half duplex.

For definitions that are relevant to the personal computer user, *full duplex* is that condition in which anything typed into the computer can be seen on the video screen. For instance, most computer systems will "echo back" anything that is typed in from a terminal. That means that if I type in something on my terminal, not only will my host computer (the one I have called) receive what I have typed, but I will also be able to see what I have typed on the screen of my monitor (Fig. 2-3). This is convenient and much safer than *half duplex*; that is the condition where only the host computer will receive what is typed in at the terminal. However, half duplex operation is especially useful for the entering of passwords, since it is very important that no one else can read a secret password when it is typed in except the receiving host computer.

On-Line

This is the state of being "logged" onto (connected into) a computer system. When you tell someone that you are having an "on-line transmission" and they give you a funny look, just tell them that you are communicating with another computer through the telephone line.

Fig. 2-3. Full duplex operation lets you see on the screen what you have typed on the keyboard.

Carrier

A *carrier signal* is a high-pitched tone that signals a computer user when he or she may go from the "talk" mode (the regular telephone) to the "data" mode (computer communications). Once the two computers are talking to each other, a carrier light will go on to indicate that everything is working properly. Don't get this term confused with *common carrier,* which is a term that refers to something like the Bell System—a company which serves the general public.

That is all of the terms that you will need to know to understand this book. However, if another one appears that you do not know the meaning of and which has not been covered here, just check the glossary in the appendices.

Ma Bell and Friends

The Bell System has become the world's most sophisticated information-handling system by steadily incorporating state-of-the-art technology that is based on the fundamental scientific advances which make high-speed digital data transmission possible. Some of the new technologies (all of which are part of what we call telecommunications) include voice-storage systems, network data bases, and lightwave communications. These new technologies will improve *the world connection* to the point that everyone will be able to take part.

Of course, there are many other things to look forward to in our telephone system, that have to do with computer communications, such as highly intelligent switching systems and high-capacity digital transmissions. The point is that today, data communications are not widespread enough to make the current telephone system obsolete. However, improved technologies, some of which are being implemented today, will carry a much heavier load of "computer talk" in the future.

Paying the Price

One misconception about computer communications is that the "free" data bases (such as bulletin board services) will not cost you anything. This seems logical, but it is not always so, because even though the data base itself costs nothing, phone costs can mount up quickly. These telephone costs may be from "message units" (the price for calls made within a certain area code but still considered nonlocal) and from long-distance phone calls.

I know that when I first received my modem, I logged onto dozens of bulletin board services all over the country, not really thinking about the cost of the telephone calls. When I received my first phone bill, though, I got a big shock. It was over $200.00 and I realized that I had done a little bit *too much* "BBSing." From then on, I learned to control my appetite for data base systems, but I would like to give some advice on how you can avoid those big bills.

First of all, you can figure that every hour of a long-distance call will cost you a minimum of $10.00 (using the low night-time rates). This is very expensive, so it might be a good idea to perhaps set a limit for yourself every month. For instance, you might say to yourself that you can call five long-distance systems during a particular month, and you will allow yourself 10 minutes for each system. A small timer placed nearby helps, or you can just set your watch alarm for ten minutes for each call. This "budgeting of time" might seem silly to some, but they are the ones who will receive the telephone bills for several hundred dollars.

Another hint is to try to stay away from the systems that are very far away, and stick with the ones that are local. Local systems are more beneficial anyway since you can develop relationships with people in your own home town.

The *real* alternative, though, is to get an alternate telephone service for long-distance calls. This can really save a large amount of money. You have probably seen advertisements for some of these services, which include *Sprint* and *MCI*. These services cost about $5.00 every month (for the service only), and if you make $25.00 or more in long-distance calls, they can be real money savers.

I use the Sprint system, which is a service of the Southern Pacific Communications Company. I know from the information I received that I can save as little as 11% to as much as 65% on my long-distance phone bills. Generally speaking, if I use Sprint wisely, I can save about 50% of my long-distance phone bill. MCI savings are comparable, and I have included the addresses and telephone numbers of both of these companies in Appendix C.

Using an alternate telephone service is very simple. I have a "password," which is a number that only I know. I also have a local access code (this is an assigned local telephone number) that I must use. Assume I am making a call to one of my favorite systems in Michigan. I would do the following:

1. Punch in my local access code (832-5016).
2. Punch in my secret password (12345678).
3. Punch in the number I want to call (313-533-0254).

The reason I say *punch in* instead of *dial* is because this money-saving service (which is perfect for computer communications) is available only with touch-tone telephones.

The Business View

Here are some interesting facts and figures to tantalize the investor who is wondering how this big boom in computer communications is going to affect money matters:

1. $250,000,000 is being invested annually by companies in a system called *"Videotex"* (a name for home information through a computer).

2. American Telephone and Telegraph, one of the giants of the corporate scene, has officially endorsed *Videotex* and announced plans for a whole new system designed specifically for data communications.
3. It is predicted that 8,000,000 homes will have computer terminals by 1990.

These facts are all certainly impressive, and it is obvious that not only are the computer buffs and futurists excited about this *world connection* news, but so are the business people. This is a positive indicator of the magnitude of this "home information revolution" as it is often called by the media.

Certainly one of the reasons for this enthusiasm is that people are in general growing more comfortable with technology. When I first ran a computer, I was a bit hesitant but, now, I don't think I could live very well without one. I have grown comfortable with it, just as many other people have grown at ease with a piece of technology that they were afraid of earlier.

AT&T is not the only major company interested in Videotex, however. Many publishers and newspapers are setting up experimental stations for things like public news broadcasting through computers and two-way television (such as the QUBE system). There are about 90 major experiments underway right now, some of which should lead to major developments in information storage and retrieval, that would benefit all of us.

As far as the newspapers are concerned, they are changing rapidly. Publishers of newspapers realize that things like AT&T's *electronic Yellow Pages* could be a big threat to their advertising sections and, consequently, their survival. Therefore, newspapers are beginning to get on the bandwagon, too, with things like electronic newspaper services and classified ad data banks. It is predicted that newspapers will not only diversify into the electronic communications media, but will also change the way that their newspapers are made. In fact, it is predicted that by 1990 publishers will be smaller and newspapers will be completely void of information like stock reports and classified ads. Therefore, it will certainly be hard to recognize a newspaper in the 1990s.

A Big Step Forward

Now that we have covered the technical jargon, the telephone service, and the business part of *The World Connection,* we can take our big step forward into examining the large time-share terminals, something that we will soon be using every day.

The next chapter will cover those large time-sharing systems, especially *The Source* and *CompuServe,* so that you can take a look at the exciting, informative, and, sometimes, funny occurrences present in these new, constantly growing, communication networks.

Chapter 3
The Big Guys

This chapter is the largest in the book because it deals with the most information concerning the "world connection," namely, *time-share systems*. These very popular systems are constantly expanding and improving, and they will probably turn out to be one of the most important mediums for all sorts of services, such as electronic mail and business news.

The most important aspects of the large time-share systems that will be covered are those that concern the personal computer user and the person in business, since those are the main functions of the time-share systems. The two key services that will be covered are named *The Source* and *CompuServe*. They have been around for several years now and will probably continue to be around for many years to come.

Defining the Networks

The giant time-share services, commonly called *networks*, are distinguishable because of two things (Fig. 3-1). The first is that they cost money to use—typically, about $5.00 per hour minimum. Secondly, these systems can support thousands of persons at one time, which is why they are called *time-share* systems. Many people can share their computer time at the same time, even though each person feels he or she, alone, is using the computer's service.

In this chapter, we are going to be examining in great detail the features of two networking services, *The Source* and *CompuServe*, along with the instructions on how to use some of the power of these two large networks. In addition, some of the "smaller big guys" who are also important to the personal computer user will be briefly described.

Fig. 3-1. Example of a network or timing-sharing system.

May the Source Be With You

First of all, we will examine *The Source*[SM] network, which is currently the second largest in the United States. *The Source* is a subsidiary of the Reader's Digest Association and is located at 1616 Anderson Road, in McLean, Virginia 22102 (Fig. 3-2). This network can be accessed by any personal computer user who has an account with *The Source*. The registration fee necessary to enroll you as a subscriber, and allow you to open a source account, costs $100.

When it was first created by the Telecomputing Corporation of America as the *first* home information retrieval system (Fig. 3-3), *The Source* promised to have on-line airline schedules, restaurant guides, and other helpful services. Unfortunately, *The Source* had a very rough start and could not offer these services when it was introduced in June of 1979. Financial problems arose, and it seemed as though *The Source* might not make it. Luckily, though, *The Source* has been upgraded, improved, and is much more heavily used today than when it was first introduced. It remains a very popular information utility.

---
*THE SOURCE is a service mark of Source Telecomputing Corp., a subsidiary of The Reader's Digest Association, Inc.

Fig. 3-2. Home of The Source (Courtesy Source Telecomputing Corp.).

Fig. 3-3. Learning while still having fun (Courtesy Source Telecomputing Corp.).

Over 20,000 customers are now using *The Source.* Becoming another one of those customers is quite easy but it can also become rather expensive. After paying the initial fee of $100, a monthly minimum fee of $10.00 is required, in addition to paying anywhere from $5.75 to $35.00 an hour for usage. (Rates depend on time of usage, baud rate of service, and whether you want standard or value-added information services.) Naturally, the higher fees are for business hours and the lower $5.75/hour cost is for those times when usage is much lower (such as after midnight). Most personal computer users *log-on* (call up the system) in the evening so that they can enjoy the lower rates, but they get the same type of service. Fig. 3-4 shows an inside view at *The Source,* where some of the giant memories used in today's computers are located.

Fig. 3-4. Giant memories of today's computers (Courtesy Source Telecomputing Corp.).

Joining the Source

The $100 one-time registration fee that *The Source* charges pays for several things, including:

1. A personal identification number (used to log-on).
2. A private password (used to log-on).
3. The local telephone number needed to access *The Source.*
4. *The Source* user's manual.
5. Subscriptions to the *Sourceworld* magazine and to the *Source Digest.*
6. A 30-day guarantee of satisfaction or your registration fee will be refunded upon written request (made within 30 days of subscribing).

These things will allow you to log-on to the network using your own secret number and password. All of your billing is charged to a credit card and you are billed monthly for the time you are on *The Source.*

There are many programs and services on *The Source.* To be more exact, there are over 1200. Naturally, I can't list all of them here, but I will show many examples of how an information utility such as *The Source* can benefit you in either personal computing or in business computing. To show how *The Source* works, I am going to have an imaginary communication *session* with *The Source* so that you can actually see what it would be like if you were to call this information utility.

An Imaginary Session

First of all, after calling up my local *Source* number, I would type in my *identification number.* This might be "ID TCA000" or something similar to that. This code is unique and it can be used both for log-on and for other people to use to send electronic mail to me. Next, I would type in my *secret password,* which might be something like "TIM."

Once these two initial procedures are done, I am ready to use *The Source.*

This network system, like most others, is "menu driven." That means that I can go to different programs and different pieces of information by selecting options from different listings (menus). For example, the first menu I would see on *The Source* after logging in would look like this:

1. Overview
2. Instructions
3. Source Menu
4. Command Level

Enter item number, or type HELP for help.

At this point, I would enter the number of the option I wanted. For example, if I wanted to go to the "Source Menu," which has almost all of the information, I would just type in the number three (3) and, then, press my ENTER or RETURN key. That is all there is to "menu driven" commands.

Suppose that I did type in a three (3). I would then see yet another menu, the Source menu, which would look like this:

1. News and Reference Resources
2. Business/Financial Markets
3. Catalog Shopping
4. Home and Leisure
5. Education and Career
6. Mail and Communications
7. Creating and Computing
8. Source Plus
Enter item number, or type HELP for help.

As you can see, with these menus, it is possible to get closer and closer to the exact category (thing) you are looking for. This main menu covers almost everything that *The Source* has to offer, so I'm going to take all eight of these options and describe what each of them can give you, one at a time.

News and Reference Resources

If you would like to have the same news that all of the television and newspapers are provided with, such as the United Press International news service, you can. You can read news stories long before they reach the evening news broadcast, and you can even stay on-line to watch the story as it develops, which is not only exciting but could also be very helpful if your line of work requires that type of up-to-the-minute information.

In addition to world and national news, local news is also available through *The Source,* along with travel and dining information, the latest happenings in government and politics, consumer information, and science and technology reports. This is a tremendous amount of information available to your call. What is even more staggering is that this information is just the "tip of the iceberg" of all the information that is available on a network like this one.

Business and Financial Markets

One area which requires constantly updated information is that of the business world. Stocks and gold prices, commodities, and interest rates are things that change and these changes can be carefully monitored through a large information utility (Fig. 3-5).

To begin with, one of the most powerful features of the financial section is the ability to keep track of 3100 stocks. The

Fig. 3-5. Work at an information utility can get very busy (Courtesy Source Telecomputing Corp.).

information about certain stocks may be organized in 12 different formats on *The Source,* which will arrange the data by certain financial attributes (such as price/earnings ratios or industry groups). This instant information can be very profitable for the investor who uses the power of a computer for profit.

Also, an analysis and computation section is available to solve your "what if" questions and to determine if certain investments would be right for you. There is also a research and reference section so you can check out those potential investments. Finally, there is a news and commentary section which furnishes various experts' opinions on the topics of the day which may influence the financial market.

Catalog Shopping

"On-line" shopping (shopping for things with a computer) is a convenience that all of us will enjoy in the future. In fact, you can enjoy it now if you are part of the *world connection* and have access to a large service like *The Source.*

One of the best things about on-line shopping is that many items may be purchased for low prices, usually wholesale. This can save hundreds of dollars. This feature is available through the Comp-U-Star buying service. With Comp-U-Star, anyone can specify generally what they are looking for (for instance, a 19-inch color tv set) and see a list of all the possibilities within seconds. From those available possibilities, a person can examine the details of each set and find out which one is best for his or her needs. There are over 30,000 name-brand items available at about 40% off retail price, so the shopping list is certainly not limited to a very small selection.

An alternative to buying new things on *The Source* is the bartering for used items. Barter Worldwide, Inc. offers their service so that you can trade something you have for something someone else has. For instance, if you have a tv set that you are no longer using and want a dishwasher instead, you could use the barter service to locate someone who needs a tv set and has a dishwasher that they really don't need. This way, the two of you can trade (barter) and you both have the things you need.

Finally, classified ads are available in the Catalog Shopping section of *The Source* that are almost exactly like the classified ads appearing in the newspapers. The advantage to these classified ads, though, as with all of the Catalog Shopping section, is that everything is electronic and instantaneous.

Home and Leisure

The Home and Leisure section of *The Source* information utility is the "fun" part of the service. Various games, such as Blackjack, Adventure, Tic-Tac-Toe, and dozens of others, can be played on the computer.

Along with the various games, there is available horoscope charts and readings, travel and dining information, and various other forms of entertainment, such as computer-generated card-tricks, puzzles, and riddles. As you can see, information services can be not only useful but also entertaining.

Education and Career

One of the great advantages of having access to an information utility is that you can gain knowledge about different career opportunities and learn some of the things you will need to know in that career. A large selection of foreign language instructional programs is available on *The Source* for such languages as Greek, Italian, and French. Along with those foreign language programs are language exercises in poetry and grammar, and there are even spelling tests.

College and career plans can be aided by using the articles on *The Source* in the education and career section. They are filled with information about financial aid for college, remedial education, management, writing as a career, and, even, operating a fast food chain. You will find that computers certainly can be helpful tools when faced with important college and career decisions.

Mail and Communications

Electronic mail is one of the things that has the greatest potential in the area of computer communications (Fig. 3-6). This type of mail is different from regular mail in that it is delivered almost instantly. For example, if you wanted to send a note to a person who also had an account with *The Source,* all you would have to do is type in the note and punch in the person's identification number. The next time that person logs on, he or she will be able to read the message right away. If you had sent the letter the usual way, it might have taken five days or more to get to that person. With electronic mail, though, the delivery time is almost instantaneous.

Another feature in the communications section is the *chat mode*. This mode will let you have a completely private conversation with another person while using a computer terminal (Fig. 3-7). Perhaps you have a friend on *The Source* whom you would like to talk to privately. By using "talk," you can do this. This may or may not be less expensive than a phone call, but there is no doubt that using a computer to talk with another person is a lot more fun, more exciting, and is very good typing practice.

Finally, there is the *post subsection* of the communications section. This is a giant classified ad section that is exclusively for Source users. Categories in the post subsection include such things for sale as computers, office equipment, and houses. Also, there is a place to complain about problems that you may have been having with *The Source* (appropriately called the *GRIPE* section), and there is yet another category if you need to find a date for Saturday night. The world of computer communications certainly has a lot of faces.

Fig. 3-6. Using a computer terminal to send and receive electronic mail.

Creating and Computing

This area of *The Source* is used for manipulating your own files, such as programs, computer data, or statistics that you have stored within *The Source* network. One of the nice things about being a programmer and a member of a network utility is that you can have the access to the computing power of a giant like *The Source* always available to you through your own personal

39

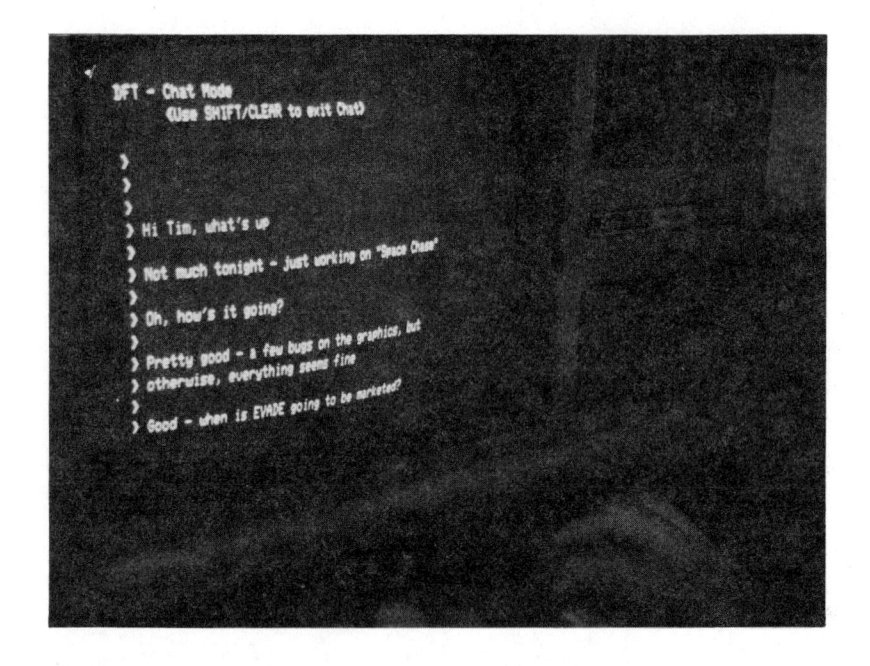

Fig. 3-7. Using the "chat" mode in the Communications section of the Source menu.

computer. This means that there is no need to buy any extra hardware or software if you want to use high-level languages like FORTRAN; languages which might have been unavailable to you before.

The storage of your own files (either programs or data) is included in your monthly bill from *The Source,* and you can manipulate these files in any way that you would like. You can sort them, check their spelling, correct their left and right margins, and even arrange them page by page. This type of convenience makes your files much easier to handle.

Source Plus

Finally, one of the newest and most interesting sections of the information service is *Source Plus.* Many business and educational journals are now available here. These include *Forbes, Harvard Business Review, Vital Speeches,* and *Venture Magazine,* among dozens of others.

Other Services

Also, there are some very special features, such as the *Mailgram* service, which is offered through the United States Postal Service. These are next-day letters which are indeed letters delivered by the post office (not electronic mail). The catch is that they cost $5.15 each. However, this may be a small price to pay for important business, and since not everyone has a Source account at this point, this service can be very useful.

Summing Up

It seems that *The Source* is on its way to becoming one of the nation's largest and most powerful information utilities. As you can see, there are some very important and helpful features offered on *The Source*. It is virtually impossible to detail all the ways that *The Source* can work for you. In fact, new programs are being added and existing services are improved continuously. I believe that you will find it worth your time to write or call *The Source* and ask for an *information pack*. This pack will help you find out if *The Source* has the things you need. There is also a customer agreement included in the pack so that you may sign up for *The Source* right away, if you wish.

A Look at CompuServe

Now, I am going to tell you what I know about the nation's largest information network, the *CompuServe Information Service* (Fig. 3-8). I use this service every day, and I think you will find some of my experiences on *CompuServe* to be intriguing and interesting, to say the least.

The CompuServe Information Service initially costs less than *The Source* since the $100 hook-up fee is not required. However, *CompuServe's* lowest rate of $5.00/hour is slightly higher than the $4.75/hour minimum of *The Source*. Considering it would take 400 hours to make up the $100 difference though, *CompuServe* can be considered a less expensive, though much larger, information service (Fig. 3-9).

However, if you think you might want to use *CompuServe,* you should write and get their information pack which has a list of the equipment that is compatible with the network, the baud rates

41

Fig. 3-8. Home of CompuServe (Courtesy CompuServe Information Service).

Fig. 3-9. Inside the CompuServe terminal room as seen through a fish-eye lens (Courtesy CompuServe Information Service).

you can operate on, and, most important, a list of the surcharges and rates charged for various services. Also included is a Subject Index listing (which is updated constantly on the line) and a list of the cities from which the service can be accessed. Write to CompuServe Information Service Division, 5000 Arlington Centre Boulevard, Post Office Box 20212, Columbus, Ohio 43220.

We are going to learn about *CompuServe* in much the same way we learned about *The Source*—by showing the initial "menu" that you would see if you called up the system and, then, going through the details of each of those menu options.

However, there will be a major difference here in that once I have finished going over all the menu options, I am going to go into greater detail about some of the unique features of *CompuServe*, such as their "Citizen's Band radio" simulation (known as CB), the Multi-Player Host, and the "Ask Aunt Nettie" column. I believe you will find these to be even more exciting than anything else I have talked about up to now. But first, let's examine what *CompuServe* has to offer in their *main menu*.

The CompuServe Subject Index

The main index of *CompuServe* is called the *CompuServe Subject Index*. It would look like this if you logged onto the information service:

CompuServe   Page CIS-1

COMPUSERVE INFORMATION SERVICE

1. Newspapers
2. Home Services
3. Business and Financial Services
4. Personal Computing Services
5. User Information

This main index will take you to any part of the information service, so we'll start our "CompuServe exploration" by looking at what each of these menu options will take you to.

Newspapers

A highlight of *CompuServe* is the number of newspapers that it supports. When on this information service, you can look through articles, weather, classified ads, or nearly anything else that you might find in a regular newspaper just by selecting the *newspapers* option on the main menu and, then, selecting the newspaper you like. These include *The Columbia Dispatch, The San Francisco Chronicle,* the *Minneapolis Star Tribune,* the *St.*

Louis Post-Dispatch, The Los Angeles Times, and the Atlanta Journal Constitution, just to name a few.

Reading newspapers through a computer can be difficult at times, however, because a newspaper article takes about 12 "screensful" of information. However, this type of access can be helpful for research and for information-gathering from a great variety of newspapers, rather than from just your local newspapers.

Home Services

Home Services constitutes a large variety of information. To begin with, a reference library is available in this section that could help you research cars, energy statistics, and new products. In addition, the research section will give you access to many government publications, timely movie reviews, predictions of the future, and even the *Better Homes and Gardens* magazine.

Home Services also can give you the power to communicate with other *CompuServe* users (Fig. 3-10) through several different ways:

1. Electronic Mail, which is almost identical to the mail service of *The Source,* allows you to send private messages to any other user of the service.
2. The National Bulletin Board, which is similar to *The Source's* "post" feature, is like a large classified ads bank. With the National Bulletin Board, you can post FOR SALE, WANTED, or NOTICE messages to everyone on the information service.
3. CB Simulation, which allows you to talk to dozens of other information service users at the same time. This will be discussed in greater detail later in the chapter.

Home Services also provides a large number of games and entertainment. These include such games as Craps, Blackjack, Civil War Simulation, and Adventure. Also, in the Home Services section is the home management computer program, which can help you balance your checkbook, calculate your net worth, and compute the amortization of a bank loan (Fig. 3-11).

Finally, the Home Services section will allow you to shop and bank at home (like the Comp-U-Star service discussed earlier in

Fig. 3-10. Using a microcomputer to check on the "sale" items listed on the Community Bulletin Board (Courtesy CompuServe Information Service).

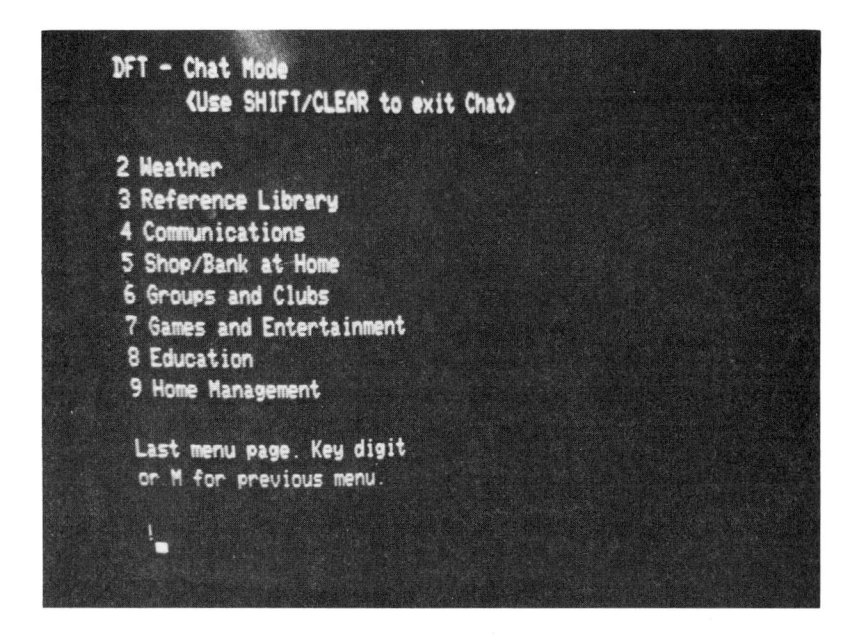

```
DFT - Chat Mode
        (Use SHIFT/CLEAR to exit Chat)

2 Weather
3 Reference Library
4 Communications
5 Shop/Bank at Home
6 Groups and Clubs
7 Games and Entertainment
8 Education
9 Home Management

Last menu page. Key digit
or M for previous menu.

!
```

Fig. 3-11. "The CompuServe Index" showing one of the many menu pages.

this chapter) and will give you access to a variety of computer-related clubs, such as the Atari computer user's group and the RCA computer user's group. Home Services offers all of these services and will continue to offer more, since it is one of the most popular selections in the CompuServe Information Service.

Business and Financial Services

For business people, the Business and Financial Services section can be of vital importance. The two main subsections are News and Reports and the Reference Databases.

The News and Reports subsection is mainly for business references such as the *Archer Commodity Report,* the *Raylux Investor's Report,* and *Investment News and Views.* Information from these references helps people in business stay abreast of happenings in the world of finance.

The Reference Databases, on the other hand, are for even more detailed information analysis, such as up-to-the-minute stock prices, current interest rates, and the *Standard and Poor's* index. The use of this information will cost a little extra money, but if you make a million dollars using your instant stock information, it would certainly justify the extra cost.

One other recently added feature to the business section of *CompuServe* is the *Lobby Letters* program. By using *Lobby Letters,* you can send a letter to any national or international elected official. By using *CompuServe,* you can create and send messages to Senators, Congressmen, and even the President. This service costs a small amount for each letter, and the cost varies for different types of letters, but this service may be a good way to get people to participate more in their own government.

Personal Computing Services

This service contains all of the computer newsletters that *CompuServe* users enjoy reading, like the Atari and the TRS-80 computer newsletters. In addition to these newsletters, there is also the current month's issue of *Popular Electronics* magazine available. This is a special service that is provided by *CompuServe* free of extra charge.

Also, a *programming and computing* section is available under the Personal Computing Services section. This section is intended for use by those persons who would like to have the power of different programming languages at their fingertips. For example, I own a TRS-80 Model III, so I have the BASIC language to use. However, if I wanted to use another language, like FORTRAN or SNOBAL, there would be no need to buy extra hardware and software for my computer since I can call up *CompuServe* and use the power of their giant computers to program in these languages.

User Information

The last index choice is *user information,* which is actually the main link between the user and the *CompuServe* network. Using this section, you can do many things, like reviewing your credit card information, finding out how much time you have been on *CompuServe* in any given week, making arrangements to change your secret password, or finding out the current *CompuServe* rates.

Also, there is a special section called FEEDBACK that will let any user give his or her comments, suggestions, or complaints directly to the network. I find FEEDBACK to be particularly helpful. The *CompuServe* people usually write back, using electronic mail, to respond to my FEEDBACK entries within a couple of days.

Extras on CompuServe

One of the minor but nice extras available on *CompuServe* is called ARTGAL, which stands for Art Gallery. While using ARTGAL, a user can browse through a catalog of computer-generated pictures and even place an order for any number of them. For instance, if I wanted a wall-sized mural of the Golden Gate Bridge, I could order it directly through *CompuServe,* and it would arrive via United Parcel Service in a few weeks. There are many pictures available, such as Mr. Spock, Alfred Einstein, and the Mona Lisa. These pictures are beautiful and are unique since they are done with a computer printer. Naturally, they cost money (usually around $7.00), but the cost will simply be added to your *CompuServe* bill.

Another extra (which is also educational) is *The Multiple Choice,* often called TMC for short. TMC is an extremely large selection of questions designed to sharpen your intellectual skills. These range from very serious scholastic questions to some more entertaining trivia questions. Analogies, word jumbles, and intelligence tests are also in TMC. It is a section that is both entertaining and rewarding. I feel that this type of feature is going to continue to improve so that one day we may receive a good deal of our knowledge solely from computer data banks.

There are some very special areas of *CompuServe* that I would like to discuss in greater detail. These are the especially *fun* sections that not only are innovative and entertaining, but are also unique to the service. These fun parts of *CompuServe* are CB, Aunt Nettie, and the Multi-Player Host.

Citizen's Band Radio

CB, as you might have guessed, is a computer simulation of the Citizen's Band radio. This feature is by far the most addictive of all the *CompuServe* programs. I, along with hundreds of others, have spent hundreds of dollars using CB.

What makes it so addictive? That is hard to say, since most people might think sitting in front of a computer "typing" to people would not be very enjoyable. Still, dozens of people stay on CB day and night, and I must admit that I am one of the many "CB-aholics."

First, let me explain how CB works. Since you cannot use voices over a computer, you must communicate in a different way. That is, obviously, typing. CB is simply a lot of persons who have logged onto the CompuServe Information Service and who are typing to each other. If I type in "Hi, everybody," everyone using CB will see my words. Anything that anyone else says will be flashed on my screen. In this way, dozens of people can talk with each other, just by typing.

There are 40 channels available on CB. Usually, the vast majority of people congregate on a single one of them. Usually, that channel is channel 1, but sometimes it is 25 or 19. There is no reason why these channels are picked but, usually, once one person is on a certain channel, the others will follow. It would make an interesting psychological study, to be sure, to examine why certain people gather on certain channels and how people commune and interact with one another through this communications medium.

Each channel has been defined by *CompuServe,* but these standards are rarely followed. Channel 1 is supposed to be the "adult" channel (though the meaning of that has still not been explained), but many younger people still use it. Other channels have amusing little names, and certain channels may be reserved for certain groups (either by *CompuServe* itself or by some friends setting a date with electronic mail). For example, a predetermined channel and time may be set for the Atari user's group meeting, or two people who know each other personally might arrange to meet on a certain channel one evening. It's an interesting way of communicating with people, which is one of the things that makes CB attractive.

Before I go into the details of CB, I would like to show you a minute's worth of CB conversation which I decided to print out. In this conversation, you will see that if anyone says anything, his or her words are preceded by something like this: *(1,Daddy Warbucks).* The information within the parentheses is the channel the person is on and, then, the person who said the words. For instance, the example above means that the words which follow were said on Channel 1 by a person whose "handle" (nickname on CB) is Daddy Warbucks. Naturally, since I was on Channel 1 at the time of this conversation, the only number in parentheses was a 1, followed by the person's handle. A few examples are: *(1,New Wave Gal), (1,Ganja), (1,Ace),* and so on. Incidentally, my handle is (and, always has been) the *Blue Knight.*

The "art of choosing a handle" is an interesting one. Personally, I picked the *Blue Knight* because my last name is Knight and because there was once a police show called "The Blue Knight" which I enjoyed. I find that there is a story behind almost every person's handle, some of them rather interesting. For example, Ms. Rainbo, one of the young and vivacious CB veterans of *CompuServe,* chose her name because the first thing she looked at when she was trying to think of a handle was her goldfish, Rainbo. She realized Rainbo was a perfect name for her (and her charming personality), so Ms. Rainbo was born. When creating a handle of your own, I recommend that you pick one that reflects something about you or your lifestyle.

The following CB conversation was picked at a completely random time, so it reflects what you will generally find on CB—a light and friendly atmosphere. Don't let any of the strange terms in this conversation throw you, since I will explain them all at the end of this minute-long CB extraction. Here, then, is a minute of CB:

(9, Blue Knight) Hi Hi Hi everybody
(9, Daddy Warbucks) Hi BK
(9, New Wave Gal) Hi Knight
(9, Ace) Hi Knight
(9, Daddy Warbucks) I am in Tulsa, Ganja, U?
(9, Blue Knight) New Wave—have we talked before?
(9, Champie Champ) Body, are you a lurker?
(9, Baud-y) Hi Dad
(9, Luna) AOS Knight
(9, Blue Knight) Hi Luna. New Wave gal????
(9, New Wave Gal) He flirts so much, he doesn't remember
(9, Ganja) I'm from Little Rock, Daddy
(9, Champie Champ) Who cares, Ganja?
(9, Ace) Who are you talking about, Gal?
(9, Daddy Warbucks) Arkansas, right, Ganja?
(9, New Wave Gal) I'm talking about Knight
(9, Blue Knight) Not nice, gal, not nice.
(9, Ace) The flirting Knight.
(9, Blue Knight) Hmmmmmmm......
(9, New Wave Gal) I was only kidding, Blue.
(9, Ace) ZZZZZZZZZZZZZZ,,,,,,,,,,,
(9, Daddy Warbucks) Y R we on this channel???

Yes, I know, it's very confusing, but after being on CB for about two years now, I've grown used to that seemingly insane confusion. However, if you look carefully, you can weed out the three conversations that are going on at one time. That is the trick to CB—people don't take turns, they just talk. There's nothing wrong with that, though, since trying to figure out just what is going on will keep your mind alert.

Some of the words and phrases in that "conversation" might be confusing to you. Therefore, I am going to give you a list of some of the CB lingo that I have acquired over the past two years. Once you understand it, you should have no problem on something like CB.

First of all, it is traditional to abbreviate everyone's name. This not only makes for a more friendly atmosphere, but also speeds up conversations. For instance, my name, Blue Knight, is often shortened to BK, Blue, or Knight. Others like Daddy Warbucks are sometimes altered even more to things like Dad, Pops, or DW.

Next, there are some words which have evolved completely from the use of CB. For example, a "lurker" is someone who says absolutely nothing on CB, but just sits at his or her terminal and watches what is going on. Lurkers are aggravating, since users would like everyone to get in on the conversation. The term "AOS" means acquisition of signal and is just a strange way of saying "hello." Also, abbreviations such as "U," "Y," and "R" can be easily figured out because they mean what they sound like—"you" and "are."

Finally, there are several pieces of CB lingo that represent other things. These include "ZZZZZZZZ" for extreme boredom (though the people who say this stay on CB anyway), and "Hmmmmmm..." for a person thinking (and who is sometimes called a hmmmmingbird by aggravated fellow CBer's).

There are a great number of advantages to using CB compared to meeting people in other ways. For instance, using CB, shy people don't have to be self-conscious or afraid of others. They can just log right onto *CompuServe* and begin talking. Also, if it's late at night and you look like you've just been through a hurricane, there's no need to worry about how you look in order to talk with your friends on CB.

Of course, CB is a great way to *make* friends also. I have made many friends using CB. One of the nice things about this communications medium is that new friends almost always have an interest in some of the things I like; namely, computers. It is not uncommon for people who meet on CB to become so interested in one another that they actually meet personally. In fact, I was recently telephoned by a friend of mine (whose handle is CompuWhiz) who said that he was in the area and we should get together. It certainly was nice meeting someone I had known for so long but never had the pleasure of meeting personally. I have done this before and I can honestly say that it certainly is a real thrill meeting someone whom you know but have never met (sort of like meeting a pen pal).

There are many commands on CB to change channels, monitor other channels, and so on. I will cover the most important CB commands here. Each of these commands is preceded by a slash symbol (/) which lets CB know that you are making a command.

51

CB Commands

The commands to CB are simple to learn and use, and CB is a tremendous amount of fun. I think that the popularity of CB will continue to grow as people realize what a wonderful way it is to meet interesting people with all types of backgrounds.

/STA—This shows how many people are on each station. For example, if you typed /STA, CB "service" might reply "(10) 5, (19) 6" which would mean that five users are on Channel 10, and six users are on Channel 19.

/MON—Monitoring channels means that you can see what is going on with another channel. For instance, if I typed /MON 6, then I would see everything that anybody says on Channel 6, in addition to the station I am presently on.

/TUN—The TUN commands will tune you into another station. If I wanted to go to Channel 6, I would just type /TUN 6.

/OFF—This command will get you out of CB and into the Multi-Player Host—something I will discuss very soon.

Life, Love, and Trivia

Aunt Nettie ("Nettie" being short for "Network") is *CompuServe's* own expert on life, love, and trivia. Aunt Nettie columns run about once every two weeks and are filled with about ten questions and answers concerning all sorts of subjects. They often range from "How can I get people to like me more?" to "Do you have a cure for baldness?" (Her answer was, "Yes – hair.")

Aunt Nettie really is an interesting sort of person. (She is indeed a person, not a program.) She answers questions as a regular columnist on *CompuServe,* and her answers are sometimes helpful, often interesting, and always amusing. Life and love are not the things Aunt Nettie handles very often, but some people do ask her questions about those two things, and she usually replies with the intelligence and wit of better-known columnists, like Ann Landers or Dear Abby. However, trivia is Aunt Nettie's strongest point, and it's a rare occasion when someone poses a trivia question that Aunt Nettie can't answer.

Aunt Nettie is a fine example of what is now an entertaining use of computer communications. Perhaps, in later years, more useful things will spring out of this, such as "on-line psychiatry," or, to a lesser degree, "on-line consultations" in business or travel. But, for now, Aunt Nettie should be appreciated for what she is—an expert on just about everything in the growing technology of the *world connection*.

The Multi-Player Host

Playing a game of *Monopoly* or *Scrabble* with friends can be a very enjoyable pastime. But, what would you think of playing an exciting Space Adventure game with many people from all different parts of the United States simultaneously? This science-fiction-sounding game is a reality today with the Multi-Player Host, the part of the CompuServe Information Service that will let many people play a game against many others at the same time.

Currently, there are two main games in the Multi-Player Host. These are *Space War* and *Decwars*. Each of these involves a futuristic space setting in which Klingons and Federation bases can fight to the death (or, at least, until they log-off the system). There are 75-by-75 sectors in *Decwars,* with half the players being on the Federation's side (the good guys) and the other half on the Klingons' side (the very bad guys). This imaginary battle between good and evil has sparked both friendships and bad feelings from user to user but, all in all, *Decwars* and *Space War* are a lot of fun and show the potential for games of the future.

For instance, the future may hold networks that have enormous strategic games in which thousands of people can play. Can you imagine the excitement when you find that every single person in Ohio is lined up with your laser turret? Some people may think that these types of games create destructive habits in people but, actually, they weaken them because of the fact that these fast-paced games release violent feelings, like anger and tension, in people. It will be interesting to see what the entertainment of the future will be like.

Summary

That pretty well covers the CompuServe Information Service. One remark I would like to make about both *The Source* and *CompuServe* is that they each have their own quarterly magazine—**Sourceworld** and **Today,** respectively. These are meant to keep users up to date with informative articles and studies.

CompuServe has a tremendous amount to offer and, for the personal computer user, I highly recommend it as one of the most entertaining and informative computer communications systems in the United States (Fig. 3-12).

Fig. 3-12. Computer networks are entertaining, informative, and have much to offer the personal computer user.

Final Words on The Big Guys

The large time-sharing systems are going to continue to get larger and will offer more and more features and power. I find this aspect of the *world connection* particularly exciting because of its great potential.

There are other smaller "big guys," such as the Dow Jones Information Retrieval Service, and many college and university computers, like the PLATO National Educational Computer System. These may be contacted from the addresses and phone numbers provided in the appendices. Each has its own advantages, but I find that the two largest time-share systems, The Source and the CompuServe Information Service, have much more to offer the personal computer user.

Now, we are going to step away from those two giants and look at the hundreds of small independently owned bulletin board services located throughout the United States.

Chapter 4
The Little Guys

Those services that I call "the little guys" are actually the hundreds of computer *bulletin board services* (BBSs) that are scattered across the United States. There are many differences between *the little guys* and *the big guys*. One of these differences is that a computer bulletin board service is owned and operated by an individual, not by some large company. In addition, BBSs are free. I will cover these differences in more detail shortly.

The reason I am going to present an entire chapter on BBSs is because they are extremely important in the *world connection* that this book talks about. There are at least 400 BBSs in the United States alone, and each of these receives about 20 to 30 calls every day of the year. BBSs are extremely popular and you will see why once I tell you about the things that they have to offer.

What Is a BBS?

A BBS is simply a personal computer that has a program in it that will allow people to call it and use the computer. A BBS needs three essential items: (1) a computer to run the BBS, (2) software to run the bulletin board service, and (3) a modem that will allow a person to call the computer and communicate with it.

Bulletin board services started appearing about 1978 and have been improving and growing in number ever since. For example, when they first started, all that BBSs could do was act as a computerized "bulletin board." A BBS could allow messages to be typed into its memory, and, then later, allow them to be retrieved. This is exactly how a regular bulletin board works. Someone posts a message for someone else to retrieve later.

Today, however, BBSs have the power that even some large time-share systems don't provide. You will see this in the imaginary "on-line session" that I am going to show next. This on-line session is what you would see on your computer screen if you called up a BBS and communicated with it.

Calling the BBS

First of all, you would have to find the number of a bulletin board service. This is simple enough, since there are hundreds of them out there. You could just check with a Radio Shack or Computerland store and ask for a BBS number. They would be happy to help you. Some Computerland stores even have BBSs of their own.

Next, while at your computer terminal, you would call that number and listen for a high-pitched whine (tone). At this point, your computer and the "host" (or BBS) computer would start communicating with one another, and you might see something like this on your terminal screen:

Welcome to Download-80 computer bulletin board

in operation since July 5, 1981

owned and operated by Preston King

Please feel free to leave any messages or comments

This type of "welcome" message would tell you what BBS you have called. It tries to give a friendly feeling to anyone who calls the BBS.

After this, you would be asked to "log-in." That is, you would have to type in your name and, then, the city and state you are calling from, like this:

What is your first name? TIM

What is your last name? KNIGHT

Where are you calling from? MORAGA, CA

You are TIM KNIGHT from MORAGA, CA, correct? YES

Once that step is completed, you would be free to do whatever you wanted to on the bulletin board service. You might want to leave a message for someone, retrieve a message that someone has left for you, or talk with the person who owns the BBS. To do any of these things, you would have to type in a one-letter command. These commands vary slightly from one BBS to another, but they usually follow a pattern (Fig. 4-1).

Fig. 4-1. The "menu" of a BBS.

The Commands

When the BBS asks you what you would like to do next, it usually will print out something on your video screen which resembles this:

Please type in the function you want: B, C, D, E, G, H, K, N, O, Q, R, S, T, U, X

Each of these letters represents a specific command, all of which are explained here.

> **B**—*Bulletins:* This command will print out the bulletins that are currently on the system. Bulletins are made by the person who owns the system (the system operator or SYSOP) and may announce things like meetings for clubs, birthdays, special events that are happening locally, and so on.
>
> **C**—*Case:* This command is used for switching from uppercase to lowercase or from lowercase to uppercase. It is known as a "case switch."
>
> **D**—*Downloading:* "Downloading" is the process of transferring

a computer program from one place to another through computer communications. For example, if you called up a system and saw a program that you wanted that was called "Word Jumble," you could use the D command to download it from the computer you have called and into your computer.

E—*Enter:* Since bulletin board services are mainly for entering and retrieving messages, this command is very important. It is used for "entering" a message into the bulletin board service. To do this, just type "E" and the computer will ask you to type your message in. Once you are through, the computer will save the message until the person you sent it to reads the message.

G—*Goodbye:* This will log you off of the computer system immediately. You cannot simply hang up on most systems, since that may cause the system that you hung up on to be unavailable to users for some time, because most systems are not designed for people to hang up on them. Always be courteous and take the few extra seconds to type **G** for "goodbye."

H—*Help:* This is a command that I used frequently when I was first introduced to bulletin boards. **H** will guide you through the system, explain its features and functions, and assist you with anything that may have you confused.

K—*Kill Message:* Not necessarily a violent thing, since "killing" a message merely removes that message from the computer's disk memory. Usually, you can only kill your own messages, since killing another's messages is not permitted on many BBSs.

N—*Nulls:* "Nulls" are simply pauses. For instance, if you had a terminal connected to a printer and you had to give the printer time to print out what was being displayed on the screen, you would have to set a certain number of "nulls" or pauses to slow down the exchange of data from your computer to the "host" or BBS computer.

O—*Other Systems:* This command will display a listing of other computer bulletin board systems. This is a nice feature to have, since you can find out about many local and national systems that you can call at some other time. For your convenience, though, I have made a listing of many of the current BBSs. They are listed in Appendix B.

Q—*Quick Summary:* If you want to look briefly at the message "headers" of the system, use the "Q" command. A message header simply gives the name of the person who left a message, who the message is to, the subject of the message, and the time and date that it was left on the bulletin board.

R—*Retrieve a Message:* This is self-explanatory, because this command will get a message for you. You can retrieve many messages, an individual message, or just the messages addressed to you. In addition, you can "selectively retrieve" messages. That is, if you wanted to see all the messages with a specific subject (for instance, "FOR SALE"), then the computer would find all of the messages on file that had that subject.

S—*Summarize Message:* This is similar to the "Q" command except that it will give you extra information, such as the length of the message, if it is protected by a password or not, and so on.

T—*Talk With SYSOP:* This command varies on many systems. It may be a "C" for "chat" or a "Y" for "Yell for SYSOP," but they all do the same thing. They summon the person who owns the system so he can talk with you (in very much the same way that you would talk with someone using CB). As you may recall, SYSOP stands for SYStem OPerator. Usually, there are specific times when the computer will let you page the SYSOP. This paging is done with a loud "beep" coming from the computer printer. However, be considerate of when you want to talk to the SYSOP, as he probably wouldn't appreciate your beeping the printer bell in the wee hours of the morning just to chat with him (Fig. 4-2).

U—*Upload:* As you might have guessed, "uploading" is just the opposite of "downloading." That is, it is the command that lets you send a program from your computer to the "host" (BBS) computer.

X—*Expert User:* Once you really think you know your bulletin boards, you can deem yourself an "expert user." This command will shorten all of the other commands. For example, instead of receiving a long listing of all of the functions you can choose from, the computer would simply ask "Function?" since it expects you to know all of them if you are an expert user.

The preceding commands represent the standard for most

Fig. 4-2. "Chatting" is not always so pleasant for the SYSOP.

BBSs, and they are all very simple to learn. Leaving and retrieving messages is easy and fun to do on BBSs, and it certainly is a lot more efficient than sending a letter.

Advantages of the BBS

There are certain definite advantages of bulletin board services over the large time-sharing systems that I discussed in the previous chapter. Even though BBSs can only support one user at a time (usually) and even though they are operated on a simple personal computer rather than a huge and sophisticated machine, the advantages balance out the disadvantages fairly well (Fig. 4-3).

The most obvious "plus" about BBSs over larger network systems is that they are free. *CompuServe* charges a minimum of $5.00/hour, and *The Source* usually charges more. Bulletin board services, on the other hand, are almost always free of charge, even though the phone call may cost money. However, there is probably a BBS within your own local area. In that case, it wouldn't cost you one extra penny.

Another advantage is that BBSs are much more personable than large time-sharing systems. If there is a problem with the computer that you are calling, you have the immediate attention of the SYSOP who will fix it as soon as he can. However, on a giant system like *CompuServe*, I have experienced long delays because of technical problems, which can become very ag-

Fig. 4-3. Do the "little guys" have a fighting chance?

gravating. In addition, friendships can develop between you and the SYSOP and, also, between you and the other users of the system.

Yet another good point for the BBSs is that there are many unique features that cannot be found on large time-sharing networks. This is because a small BBS can easily fit the needs of the 30 or 40 people who call it regularly. In fact, suggestions from users make BBSs almost "custom designed" to their users. On something like *The Source,* with such a variety of users, it is impossible to meet the needs of everyone, so you will not find the "frills" like uploading and downloading games, local happenings, and chatting with the SYSOP that you will find on most bulletin boards.

Bulletin board services certainly do have advantages of their own. There are many different kinds of BBSs, too, and finding out their unique features is always exciting. I will be discussing the different types of BBSs (such as the FORUM-80, the ABBS, and the Bullet-80) in Chapter 7.

Using a BBS Wisely

You can make a computer bulletin board service one of your most valuable tools, if you know how to use it wisely. There are three main steps to doing this:

1. Get to know the system.
2. Get to know the SYSOP.
3. Get to know the other users on the system.

Getting to know the system is simple enough. By using the "H" command, you can find out what advice the bulletin board itself has to offer. Use each of the commands to find out what they will do, and read any bulletins or special notices from the SYSOP about the system. The best teacher is experience, and you will soon find that using a BBS is not as hard as it may seem.

The SYSOP is your next step. Use the chat command, whatever it may be ("C", "S", or "Y"), to call the SYSOP for a little talk. Tell him about yourself, your interests, and comment on how much you appreciate the system. This type of talk will help you and the SYSOP become friends. You may find that the system operator can offer some good advice to you on almost any computer-related subject.

I have met many SYSOPs and a great number of them have become my friends whom I write to on a regular basis. We can exchange ideas, programs we have made, and advice. Meeting a SYSOP can be a rewarding experience and if you are going to be using a system regularly, it is important that you have at least *one* friendly conversation with your system operator.

Finally, get to know the users of the system. Leave messages to one or two of them to introduce yourself and you will find that a "chain of friends" will quickly build up. It is true that the more people you know on a system, the more recognized and assisted you will become. People are almost always willing to help each

63

other out, and BBSs are also a great way to make a lot of new friends who have interests that are the same as your own.

Future of the Bulletin Board Service

Bulletin board services currently are offering an efficient communications medium for a small number of users. There are new features appearing on BBSs nearly every week, which include the following:

1. Extensive downloading capability. Some BBSs have a lot of memory available, so they can hold many, many programs for people to "download" (pull from the BBS system and put into their own computer's memory). One BBS in New York has about 600 programs ready for downloading, though most BBSs only have a few dozen.
2. Product ordering. Just like with Comp-U-Star, you can order things via a computer terminal. However, there will not be as wide a selection as on Comp-U-Star, but you will probably receive the package much faster since you are calling a small BBS. Perhaps you can even negotiate the price down if you are a friend of the SYSOP.
3. Multiuser capability. Though not available on many systems, some BBSs can support more than one caller (perhaps five or six). However, because of the expensive hardware required to do this, many SYSOPs do not offer this feature. Still, there should be more of these free multiuser BBSs across the country as hardware prices go down and the popularity of owning a BBS increases.

Bulletin board services are definitely here to stay, because they benefit both the SYSOP and the users of the system. I find BBSs exciting, informative, and a great way to make friends. I recommend that you give BBSs a try. I'm sure that you'll be hooked on them from that day on.

Chapter 5
Naughty, Naughty—An Exposé

It is an unfortunate fact that many things that are intended for good purposes can be changed by individuals and used for malicious or selfish purposes. For example, in the Great Depression, President Franklin Roosevelt helped bring Social Security into existence, and today it serves the needs of the disabled, the aged, and the unemployed. However, criminal individuals have twisted this well-meaning program into something negative by extracting large amounts of money from the program through fraud. This is just one example of how something "good" can be turned into something "bad."

Another good example is that of video cassette recorders, or VCRs, as they are often called. These devices can be used for entertainment, for recording various events, and for photographing valuables for insurance purposes. Still, there are many who use this modern miracle for their own illegal purposes, such as copying current movies from pay-television and, then, selling these copies for a profit.

Computers are no exception to this "using good for evil" idea, for the clever criminal knows that he must keep up with technology. Computer communications is one of the newest mediums for "computer crime" and its associated actions of embezzling money, altering data, and cheating others.

Examples of Computer Crime

There are countless examples of computer crime that has been conducted both by professionals and by amateurs. Many of these rip-offs are not made using some brilliant system found in a detective story, but rather by using very simple techniques to fool computers. For instance, one young man applied for a 12-month installment loan from a

New York bank several years ago. The bank sent him the loan, along with 12 computer-coded coupons which he was supposed to send in each month along with his payment. One coupon for January, one for February, and so on. Knowing something about computers, he made out a check for one payment (about 8% of the loan), sent it in along with the last coupon (meant for his last payment), and waited. Sure enough, he received a note from the bank, praising him for paying off his entire loan so promptly and assuring him of his excellent credit standing.

Not all operations are as incredibly simple as that. Other computer crime schemes involve secret midnight break-ins at computer centers, illegal exchanging of passwords, and other things that we usually associate with the standard "cloak and dagger" type of story. Computer crime, however, has developed into a professional pastime for some individuals, and it will certainly continue to grow all around us as time goes on and technology grows even more sophisticated.

The Home Computer Criminal

The rest of this chapter will be concerned not with the billions of dollars made in the different types of "computer crimes" that I have just mentioned, but will, instead, focus on the illegal acts of many *personal computer* users. The two main areas of home computer crime are in "software piracy" and "phone phreaking."

I have researched both of these personal computer crime areas through news articles, case studies, and through personal observation of what I have frequently seen on bulletin boards and time-sharing systems across the United States. If you are a part of the *world connection,* or if you intend to become a part of it, you will almost certainly run across some type of computer crime. Because of this, I think you will find my "investigative reports" on phone phreaking and software pirating interesting.

Yo Ho Ho

The "pirating" of a few centuries ago seems, to many of us, a romantic adventure of the past. However, there is a different kind of pirating that is not romantic, but merely illegal. It is called "software pirating" and has reached such great proportions that even the Congress of the United States is debating bills that would discourage software piracy.

To understand software piracy, you must understand how a program is manufactured and marketed (Fig. 5-1). First of all, the program has to be made. This is done most of the time by an individual called a *freelance programmer* (like myself) who writes a program and, then, submits it to a company for consideration and possible marketing.

Fig. 5-1. Software piracy can be at any age level if a computer is available.

If that program is accepted, it is published electronically or otherwise, and the program is then marketed nationwide, either through stores, by mail order, or both. Prices on programs vary greatly, but for the sake of this explanation, we will use the typical price for a game program—$19.95. Once people buy this program, it is agreed that the person who buys it may only use it on his system and that he will make copies only for his own personal use and protection in case something should happen to the original copy.

This is where piracy comes in. If the person who bought a program made a copy of it and gave it to a friend of his, he would then be guilty of a violation of agreement and copyright infringement, along with several other offenses. He could also be sent to prison and fined $50,000. He is guilty, but is he caught? *Of course not.* This type of "giving a program to a friend" exchange occurs thousands of times, every single day of the year. If the copyright laws were strictly upheld, almost every home computer owner might be in prison.

Why aren't these pirates arrested and convicted? Well, how is the law going to catch them? It is all too simple to make a copy of a program, and then give it to a friend. For that matter, it's very simple to make 100 copies for 100 friends. It's even easy to make 1000 copies, and then sell them to 1000 people whom you don't even know. It happens frequently, and it has been estimated by some sources that for every program sold, five illegal copies are distributed to other people.

You may be asking yourself, "How does this hurt anybody? There is nothing wrong with giving a friend something." Well, that is how many people think, including myself until a short time ago when I started having my own programs marketed. The two parties that this "exchange of programs" hurts are the programmer and the company marketing the program. The reason it hurts these two parties is because no money is paid for the programs traded, and the programmer and the company are cheated out of money that they would have received had everyone bought the program legally. Instead the majority of the people are receiving it for free.

Now you may be thinking, "The people trading programs probably wouldn't have bought them anyway, so how do the programmer and company lose money?" This is also a frequently asked question. True, not all the people who receive pirated programs would have bought them, but some would have. Also, it is ethically and morally wrong to trade programs because a pirated program represents weeks and weeks of work by a programmer. To simply give it to a person for nothing (or even for a very small price) is, in effect, stating that the program is worthless while, in actuality, it may be worth a great deal.

What does pirating have to do with computer communications? A great deal. This is because using modems is probably one of the best ways there is to pirate programs. Trading programs through the *world connection* is something that is done frequently and habitually.

How It's Done

Before I explain exactly how software piracy, through the use of modems, is done, I would like to state that I do not recommend piracy. It is illegal, it hurts both the programmer and the company that made the program, and if you are caught, you could be in very serious trouble. I am telling you how the piracy occurs only to show some negative points of the *world connection*.

One of the most common and simple ways of pirating a program through a modem is to simply become friends with someone, say on a BBS or a time-sharing system, and ask if they would like to trade programs or not. If they say "yes," all you would have to do is call them up, ask them for a particular program that they have, and then "download" it from them.

Another pirating method that is often used through modems is contacting a BBS. There are several bulletin board services that have copyrighted software available (often called *pirate boards*). Note that not all software on BBSs is illegal. Usually, a BBS has only "public domain" software; i.e., software that is not copyrighted and which can be traded by any individual. "Pirate boards" are the ones that have copyrighted programs for downloading. Downloading these programs is just as simple as downloading any other program, so the pirating of programs is almost commonplace for the SYSOPs and users of these BBSs.

Understandably, most of the pirate boards that exist today are very secretive about the fact that they have copyrighted software. The SYSOPs of these systems have passwords to protect the sections of the system that offer pirated software, and they give the passwords out only to those individuals they know and whom they can trust to keep their mouths shut.

A few systems, though, are very open about their pirating and have such names as "The Pirate's Cove BBS" and "Pirate's

Harbor." The Pirate's Harbor was started by a person who calls himself "Red Rebel." He wanted to start up a system that would get a lot of use, and it does. Since people do not want to "log-in" with their real names on a pirate board, they use creations such as "Blackbeard," "Mr. Xerox," and "Disk Zapper" to hide their true identity, but still show that they enjoy pirating software.

The people who call these systems are not green-eyed villains who despise progammers and feel only for themselves. These are people just like you and me. Some of them are presidents of giant corporations, lawyers, and stockbrokers. However, they are all breaking the law by participating in the pirating game.

Some pirating systems are so open about their activities that they hold special "Pirate Fests" near computer shows, which always draw a great number of people. These festivals include seminars on how to copy software effectively, how to pirate programs for fun and profit, and how to "crack" (find the password for) computer programs that have been password-protected by their manufacturers. Of course, these "Pirate Fests" are not met with wholehearted approval by the software manufacturers, and there was a rumor going around that one software manufacturer shook a pirate by the neck, in frustration and aggravation, at one of the pirate parties.

The last way to trade programs through computer communications is indirectly. That is, a person might put up a message on a BBS or time-share system announcing that a computer users' group will be meeting at a certain time and place, and that "special programs" will be available for the taking. "Special programs" are illegal copyrighted programs. This indirect way of trading programs by means of a modem is the third and final way to pirate programs in the *world connection.*

What Is Being Done To Stop Piracy

Software makers are not ignoring the fact that many programs are being pirated, and they are doing a variety of things to protect their programs from illegal distribution.

1. *"Protecting" the programs*—This phrase almost always means scrambling the data on a computer diskette so that it may not be directly backed up onto another diskette for illegal distribution. Still, there is a saying that "any code devised by man can be broken by man," so a lot of *software*

cracking goes on with these so-called "protected programs." Software cracking, as I mentioned, is making a protected program "unprotected." Therefore, this type of protection is really not a protection, but just extra temptation to pirate a program since it is that much more enjoyable to figure out how to illegally copy the disk.
2. *Giving computer programs serial numbers*—Serial numbers on programs are just like serial numbers everywhere else; they are unique numbers assigned to individual people. This is a deterrent to pirating for many people. For example, if I bought a program with the serial number "123" and I began pirating it, the company that made the program might find out about it. All they would have to do is check the number of the pirated program that they had found—"123"—and they would have me. However, this is not a foolproof method, either, since the registration number can also be easily altered, or even deleted.
3. *Documentation*—The documentation, or instructions, of some programs is vital. By having extensive documentation only within a written manual and nowhere within the software program, a software company can keep piracy down. This is because not many pirates want a program they don't know how to use. Once again, though, there is a way around this for pirates; merely copy the documentation. However, this can become expensive for some of the 150-page manuals that exist.

Piracy is definitely a major computer problem, and no one knows when or if it will ever end. Modems are a perfect way of distributing illegal software undetected, and it is unfortunate that something intended for "good" has been made into a device for illegal intentions. Stiffer laws are being made against pirates, though, and a current bill that is in Congress imposes a $250,000 fine on someone caught in the act of piracy. This should certainly make a person think twice about giving a copy of a program away, so perhaps these stronger laws will remove some of the piracy problem.

Phone Phreaking

Phone Phreaking is basically the ripping off of the telephone system. It is the illegal act of gaining certain codes so that a per-

Fig. 5-2. Phone "phreaks," BEWARE!!

son may make any number of calls to any place in the world at any time for absolutely no charge (Fig. 5-2). (The only reason, however, for the spelling of "phone phreak" in such a manner is merely because phone begins with "ph.")

There are a number of ways used to avoid charges on telephone calls in the "phreaking" process. One of them is to use an alternate phone service, such as SPRINT and MCI, except with illegal codes. Using a "secret code" is all too easy, and there are many people who have these codes and who use them frequently. However, SPRINT and MCI are aware of this, and they are cracking down on phreaks who are using illegal access codes. By tracing down the telephone number of the person making the call, they can pinpoint those who are using their system illegally.

However, true phreaks have found a way around this. By using special "boxes," they can avoid being traced altogether. These boxes, called "bleeder boxes" or "blue boxes," are inexpensive to build, and are being made by many phreaks for profit. These boxes are also illegal, so a person who is caught "phreaking with a bleeder box" will be guilty of several crimes.

Phreaks are often young people who are intelligent, but bored with their life, so they create havoc among telephone system computers, in addition to computers that are owned by other companies. For instance, in 1980, four 13-year-olds from a private school in New York City used their school computer to gain access to the computers of several corporations. How they found the passwords, I don't know, but they might have discovered them through some friends or merely through a trial-and-error process. Nevertheless, they erased about 10,000,000 pieces of data over six months time and juggled the huge accounts of several companies. They were caught, but, however, their only punishment was chastisement. They did promise not to do it again.

Many times, phreaks grow bored even with stealing telephone access numbers and corporate passwords, so they go onto even more complex things. Through a story in the local news, I learned about a 16-year-old boy, living near me, who knows the password to the ARPANET computer that is maintained by the United States Defense Department. This is a $3.3-million computer that is intended for use only by Defense Department contractors, but he has, along with hundreds of other unauthorized users, used the giant computer as a dating service, a pen-pal club, and an electronic magazine for youngsters. It is almost funny to think that young people all over the nation have outsmarted supposedly intelligent computer controllers, though, in a way, it is frightening, since your bank account or personal information could be examined, altered, or distributed by some 14-year-old kid with a computer who is bored with his current lifestyle.

The ultimate challenge for a phone phreak is to break into a computer-switching station for the telephone network. Recently, several youths in Los Angeles did this very thing, and stole manuals, password books, altered data on the computers, and fooled the guards into thinking they were employees of the telephone service. They gained the power to turn entire telephone systems on and off, tap the phone lines, gain any computer program they desired, and they stole $300,000 of telephone equipment. They were caught and convicted, however, but many others get away with similar activities.

Basically, the phone phreak is someone who is either out to avoid costly telephone bills or who is just looking for a good time. They are intelligent and knowledgeable. It is unfortunate that they are using their talents to destructive ends. Stricter laws and effective tracking methods are taking their toll on phone phreaks, though, and the phreak may very well become a thing of the past.

Where Will the Naughtiness End?

As technology improves, new activities in computer crime will surely develop, probably just as the old problems are being solved. The criminal element will always be there, and the *world connection* is serving his needs very well.

No one knows if these types of illegal activities will ever be brought to a halt, because of the difficulty in tracking down people like pirates and phone phreaks. Still, we must do everything in our power to make something like computer communications a positive factor in every possible aspect. We must work creatively to eliminate the negative elements that are present in the system today.

Chapter 6
A Multitude of Modems

Purchasing a modem can be a very confusing ordeal. There are many factors to consider, such as what price is best for you, what features you need, and is the modem compatible with your brand of computer.

In this chapter, the opinions of many people are given so that you can use these opinions as a "buyer's guide to modems." In this way, you can find out which one might be best for you. In addition, I have also included a section of stand-alone terminals (those which require no computer) and have listed some of their features.

Terms To Look For

To refresh your memory and to help you in understanding how some features of a modem could affect you, I will briefly go over some of the computer communications terms again. As you read the term, decide if you would need such a feature or not. This will greatly affect your final purchase of a modem.

> *Auto-Answer*—A modem that has this feature will automatically answer the phone for you if it is called. You will need this feature only if you are planning to open up a bulletin board service, for, in that case, you will be getting many calls. If you are not going to open up a BBS, then the auto-answer feature is probably not important to you.
> *Auto-Dial*—Some modems will automatically dial numbers for you. This may be helpful if you have only a few numbers that you dial frequently, but this feature is usually unnecessary.
> *Acoustic and Direct Connect*—These are the two basic types of modems. An acoustic modem is the type where you press the telephone handset into the modem and it is usually less expensive. A direct-connect modem is

more expensive, but there is no need to push the handset into the modem because it is electrically connected (via wires) to the telephone terminal box. With this method, you will have a more reliable data transfer.

The preceding definitions are the basic terms that you will need to know about when buying a modem. Now that you "understand" them, you can take a look at some of the following opinions on modems. These opinions are in the form of "reviews" (something like book or movie reviews) so that you can look at both the features and shortcomings of the modem which you may be considering to buy. I have, as with everything else, included the addresses and telephone numbers of some of the modem and terminal manufacturers in the appendices.

One last thing I would like to mention, though, is that many of these modems are for the TRS-80® Model I or Model III microcomputer. This is because a large number of modems are made solely for those microcomputers, while modems for other computers are usually made by their manufacturer. For instance, if you had an ATARI®, then you would get the modem that ATARI® sells. Still, I have tried to include every major modem in these reviews so that you can make an honest judgment about almost any major modem.

The LYNX®

Emtrol Systems, Inc. introduced the LYNX in the 1980s for three types of computers—the TRS-80®, the Apple II®, and the "RS-232" version. The TRS-80 and Apple versions work for their respective computers, while the "RS-232" version works with almost any computer that has an RS-232 "bus" or connector. The modems are all very much alike, except for their colors.

One of the main features of the TRS-80-compatible LYNX is that it does not require any special connectors, such as the "RS-232" card or the "expansion interface." Both of these are costly items. Instead, LYNX will work on any TRS-80 Model I or III microcomputer, no matter what connectors or adapters are avail-

---

TRS-80 is the registered trademark of Radio Shack, Ft. Worth, TX 76102.
ATARI is the registered trademark of ATARI, Inc., Sunnyvale, CA 94086.
Apple and Apple II are the registered trademarks of Apple Computer Inc., Cupertino, CA 95014.
LYNX is the registered trademark of Emtrol Systems, Inc., Lancaster, PA 17602.

Fig. 6-1. A telephone linkage system for the TRS-80® microcomputer (Courtesy Emtrol Systems, Inc.).

able. This, along with its relatively modest price of about $300 (though you can get them for about $220 from some mail-order companies), makes it an interesting prospect for many TRS-80 owners. In addition, it is an excellent unit for Apple and RS-232 computer owners.

The LYNX, itself, is a direct-connect modem. It communicates at 300 baud (Bell 103 standard), just as all of these modems and terminals do. In addition, it is very simple to use as there are only two main switches. The first one of these, the DATA/TALK switch, is the most important one. If you want to use your phone just to talk to someone, you will be in the TALK mode. However, if you want to use your computer to communicate, you will flip the switch to DATA. The second switch, located on the back of the unit, is the ORIGINATE/ANSWER switch. This switch will allow you to either communicate regularly (ORIGINATE mode) or will set the modem to answer the phone (ANSWER mode).

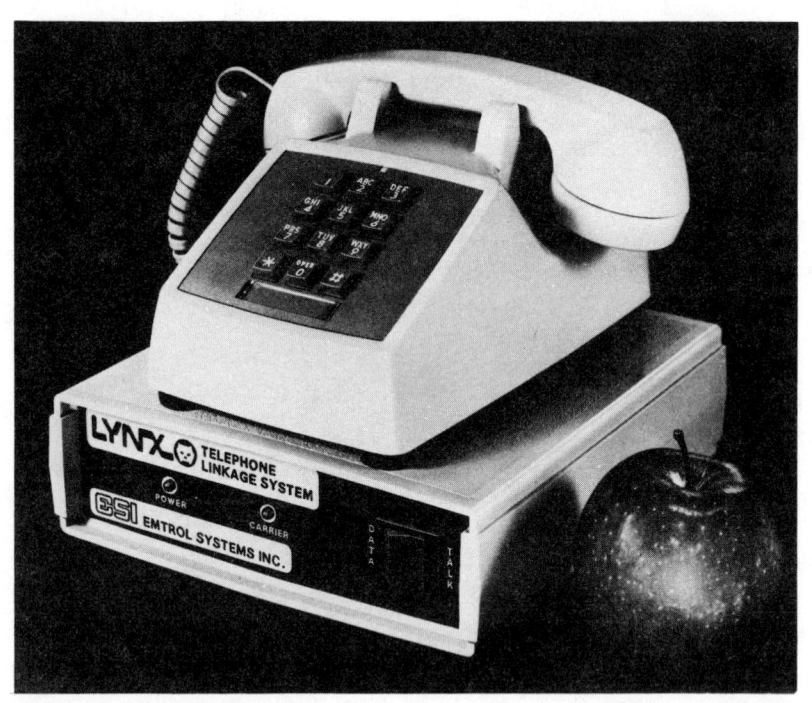

Fig. 6-2. A LYNX designed for use with the Apple® microcomputer (Courtesy Emtrol Systems, Inc.).

The LYNX has its own power supply that uses 115-volt ac 60-Hz line voltage with a plug-in adapter that is designed to supply either 135 mA or 230 mA at 12-volts dc (depending on the unit you have). Two indicator lights on the front panel of the LYNX will tell you two things: (1) If the LYNX is "on" or not (it usually is "on" all of the time), and (2) if you are currently communicating with another computer system. Also, the LYNX has an automatic dial/automatic answer capability that enables your computer to place and answer calls on its own, according to your preprogrammed instructions. The indicator lights will flash green and red when you are dialing (which is always fun to watch).

Finally, the LYNX comes with its own software "package" called EMTERM. The EMTERM terminal program is supplied on cassette; it is easily transferred to disk. This is a very basic package, for it contains only a host program, a simple terminal program, a list of on-line computer phone numbers, and some rather skinny documentation. Still, it is enough to get you started in trying out your LYNX.

Advantages of the LYNX

1. Economically priced at $220 to $300. Also there is no need to buy any RS-232 boards, expansion interfaces, or any other modifications.
2. Different versions are available for the Apple II, the TRS-80, and the RS-232 microcomputer.
3. It features auto-answer, auto-dial, simple operation, and is software-compatible with other computers. Works with any software, including ST-80 by Micklus.

Disadvantages of the LYNX

1. Very poor documentation—only a few pages.
2. More bulky than most modems.

Hayes Smartmodem®

The Hayes Stack™ Smartmodem® is so named because it is an "intelligent" modem and the exclusive Hayes Stack™ design ends clutter by letting you stack other Hayes components right on top of your Smartmodem®. The Hayes Smartmodem contains its own built-in microprocessor, highly advanced electronics, and it can be used effectively even with very "dumb" software, because it has its own internal programming. It can be controlled using any programming language. Over 30 different commands can be written into your programs or entered directly from your keyboard.

The Smartmodem (Fig. 6-3) provides many capabilities, but it must be used with a computer that has an RS-232 board, just like most modems. The Smartmodem can answer the phone, dial a number, receive and transmit data, and then hang up the phone—all automatically. In fact, it is a very sophisticated full or half duplex 300-baud originate/answer modem. This modem costs around $300, but it can be bought for a much lower price through some mail-order firms.

Smartmodem, Smartmodem 300, Smartmodem 1200, and Hayes Stack are trademarks and registered trademarks of Hayes Microcomputer Products, Inc., Norcross, GA 30092.

Fig. 6-3. Hayes Smartmodem® (Courtesy Hayes Microcomputer Products, Inc.).

Some of the best features of the Smartmodem are not found in most modems. For instance, there is a speaker in the Smartmodem which allows you to actually *hear* the computers communicating with one another. This is rather cacophonous, but it can be useful if you want to listen to know when both computers are connected and communicating properly. In addition, the modem itself has seven LED status indicators on its panel, which indicate if the auto-answer is on or off, if a carrier tone is present, if the phone is off the hook, if you are receiving data, if you are transmitting data, if the computer terminal itself is ready, and, finally, if the modem is ready.

Along with all of these features, the modem monitors the phone line for special commands. These commands are preprogrammed within the Smartmodem's Z-80® microprocessor. There are commands for Touch-Tone® dialing or pulse dialing (or a combination of both), for changing the baud rate (from 0-300 baud), and for full and half duplex switching.

The Smartmodem is easy to use because it does almost everything for you. It dials a number (on any phone system), shows the status of the phone call, and allows the software to take over once

---
Touch-Tone is the registered trademark of American Telephone and Telegraph.
Z-80 is the registered trademark of Zilog, Inc.

everything is accomplished. The Hayes Stack Smartmodem comes highly recommended to me, and with all of these features, you might want to take a serious look at it.

Hayes Smartmodem Advantages
1. Has direct-connection, auto-answer, auto-dial features.
2. Has a speaker for audio monitoring of communications.
3. Has a two-year warranty and self-testing command.
4. Has 7 indicator lights, an RS-232C interface, and includes several switch-selectable features that let you tailor performance to your exact needs. You can "set it and forget it" for the ultimate in convenience.
5. Is "intelligent"; it has its own microprocessor. Over 30 different commands can be written into your programs or entered from the keyboard.

Hayes Smartmodem Disadvantages
1. No actual disadvantages that I know of, but a modem this advanced may cause some buyers to think it is too much for them, even though this is not the case.

Hayes Smartmodem 1200®

The Hayes Smartmodem is now called the Smartmodem 300 since Hayes has come out with the Smartmodem 1200 (Fig. 6-4). The Smartmodem 1200, which is four times faster than any 300-bps modem, allows *any* computer with an RS-232C connection, such as the IBM Personal Computer, the TRS-80®, or the Apple III®, to communicate over telephone lines with other computer terminals or printers. The Smartmodem 1200 connects directly to any standard telephone jack in the United States.

Like the Smartmodem 300, dialing is by Touch-Tone® or pulse (or both), an internal speaker lets you hear the call being made and monitors its progress, and indicator lights keep you posted on the operating status. It can even operate over multiline telephone systems (PBX) to dial numbers, transmit and receive data, and disconnect—all automatically.

TRS-80 is a trademark of Radio Shack, Ft. Worth, TX 76102.
Apple III is a trademark of Apple Computer Inc., Cupertino, CA 95014.
Touch-Tone is a trademark of American Telephone and Telegraph.
The Source is a servicemark of the Source Telecomputing Corp., McLean, VA 22102.

Fig. 6-4. Hayes Smartmodem 1200® (Courtesy Hayes Microcomputer Products, Inc.).

Smartmodem 1200 is two modems in one. Like the original Hayes Smartmodem, it can communicate with other Bell 103-type modems at up to 300 bps, plus it can operate as a 1200-bps modem for communicating with the faster Bell 212A-type modems. Unlike many other 1200-bps modems, Smartmodem 1200 lets you select either full or half duplex operation for compatibility with time-sharing services or any other system you might choose. Smartmodem 1200 allows you to access The Source[SM], communicate with your branch offices, exchange programs with other computer users, or perform just about any communication function that you can imagine, and it can be program controlled using *any* language.

Hayes Micromodem II®

From the same company that makes the Smartmodem comes the Hayes Micromodem II®, a data communications system for the Apple II and the Bell & Howell computers (Fig. 6-5). This system consists of three parts:

1. The printed-circuit board, which plugs directly into the computer.
2. The Microcoupler[TM], which connects the computer directly to the telephone line.

Micromodem II and Microcoupler are trademarks of Hayes Microcomputer Products, Inc., Norcross, GA 30092.

Fig. 6-5. Hayes Micromodem II® and its terminal program (Courtesy Hayes Microcomputer Products, Inc.).

3. The software, which consists of 12 different programs for computer communications.

This system is convenient because it is easy to install, is very compact, and is easy to use. The Micromodem II has automatic dialing and automatic answering features, which are a pleasant surprise in a modem this compact. The Micromodem II also sells for the standard modem price of about $250. It is certainly a good bargain for those Apple II or Bell & Howell computer users who want something compact with good features and good software.

The printed-circuit board fits, quickly and easily, into your Apple II, eliminating the need for a serial interface card, and the Microcoupler™, which is included, connects the Apple II directly to a standard modular telephone jack. The auto-dial and auto-answer features are built into the Micromodem II® and its operation can be either full or half duplex, with a transmission rate of 300 bps. Also, it is Bell 103 compatible.

Micromodem II® Advantages

1. It is compact and easy to install.

2. Software is included, along with good documentation.
3. It is a direct-connect modem, with auto-answer and auto-dial features.

Micromodem II® Disadvantages

1. No serious disadvantages. This is a good modem buy for the Apple II user.

Hayes Terminal Program

Software is sold both with the Micromodem II® and separately. A new Terminal Program developed by Hayes specifically for the Micromodem II® allows you to access all the features of your modem within seconds and use your CP/M®, DOS 3.3, or Pascal formatted disks to create, list, or delete files. The Hayes Terminal Program is a complete stand-alone disk and, because it is menu driven, you can set your communications parameters and change hardware configuration directly from the keyboard. The Program even allows you to generate ASCII characters that normally are not available from Apple keyboards. This further extends your capabilities. A Terminal Program disk and user manual now comes with the Micromodem II®, but if you already have the modem, you can buy the Terminal Program separately.

Telephone Interface II

The Telephone Interface II (Fig. 6-6) is an acoustic modem that is available from Radio Shack, a division of Tandy Corporation. This modem will only work on the TRS-80 Model I or Model III microcomputer and, to get it to run on them, the computer must also have an RS-232C interface board that costs about $100 at the Radio Shack. For Model I users, there is an extra $300 charge because you will also need an expansion interface in order to make the modem work. As you can see, a $200 modem like the Telephone Interface II can be just a little more expensive than the price tag shows.

One of the reasons that this is a less expensive modem is because it is of the acoustic type. That means that when you call up a system, you have to listen for a signalling tone (a high-pitched whine). When you hear the "whine," you have to press the

---

CP/M is a registered trademark of Digital Research, Inc.

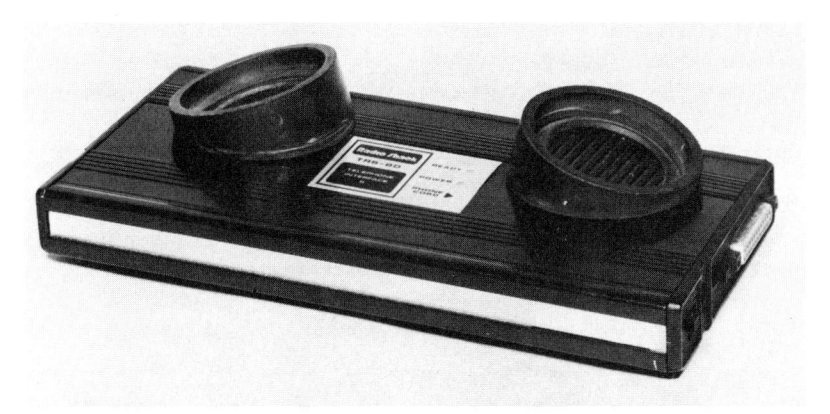

Fig. 6-6. An acoustic coupling modem from Radio Shack.

handset of the phone into two rubber cups in order to make an acoustical coupling and allow the modem to work properly. The Telephone Interface II works well, though, as long as there isn't too much external noise that it has to shield out.

Advantage of the Interface II

It is less expensive, if you already have an RS-232C interface board (and, for Model I owners, an expansion interface).

Disadvantage of the Interface II

This is an acoustic modem, so any loud noise or an accidental bumping of the modem could ruin an entire data transmission.

Direct-Connect Modems I and II

These two modems, both made by Radio Shack, are of the direct-connect type and work on both the TRS-80 Model I and Model III microcomputers. The differences between the two modems are their prices and their features.

The Modem I (Fig. 6-7) costs $149, which is less than the Telephone Interface II. It can answer calls, and it is very simple to use. Another good thing about the Modem I is that it *will* work on a Model I microcomputer without an expansion interface, as long as you buy the special cable and software sold by Radio Shack for a total of $14.90.

The Modem I is an originate/answer modem that can be directly connected to a telephone line. It will provide full duplex

Fig. 6-7. Direct-Connect Modem I from Radio Shack.

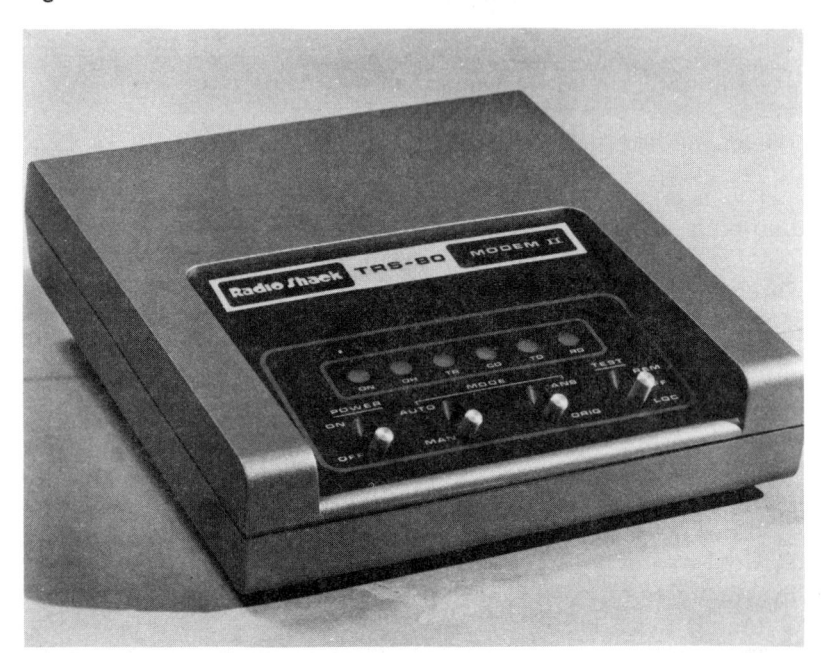

Fig. 6-8. Direct-Connect Modem II from Radio Shack (Courtesy Radio Shack, a Division of Tandy Corp.).

operation when used with RS-232 equipped TRS-80 computers, and will permit half duplex operation on Model I systems when it is used with the optional software and cable that was just mentioned.

For an extra $100, you could purchase the Direct-Connect Modem II (Fig. 6-8), a fully programmable direct-connect modem with automatic answering and automatic dialing capability and both remote and local test modes. It will automatically dial or answer the phone, receive or transmit data, and will even hang up the phone. Its baud rate is 300 baud and it will work on either a rotary or tone dialing system. There are several LED light indicators to show when the terminal is ready, when a carrier is detected, when there is transmission of data, when there is reception of data, and when the phone on/off hook is activated. This modem became available in late 1981, and it is certainly Radio Shack's best modem for the Model I or Model III TRS-80 microcomputers.

Advantages of Modems I and II

1. Has direct connection and answering features.
2. Has automatic dialing (Modem II only).
3. Has LED indicator lights that show operational status.
4. Are easy to learn how to use.

Disadvantages of Modems I and II

1. Requires an RS-232 interface in order to operate.
2. Are not for multiline use.

Modems From Novation, Inc.

The CAT^{TM} modem from Novation is an acoustic modem (Fig. 6-9) that is often considered to be one of the very best modems around. It is "sleek, silent, and responsive" as the people from Novation call it. That is, it is "sleek in styling, silent in performance, and responsive to your needs."

The CAT^{TM} is all of these things and more. It can operate from 0-300 baud, and it has answer, originate, and test modes. It can be operated either full or half duplex. Because of its compact power pack which plugs directly into wall sockets, the CAT is also

---

CAT and D-CAT are trademarks of Novation, Inc., Tarzana, CA 91356.

Fig. 6-9. The CAT™ acoustic modem (Courtesy Novation, Inc.).

designed to reduce heat and voltage hazards in the unit. A CAT can communicate with any Bell 103-compatible modem.

The CAT's price of $189 is also attractive to most prospective buyers. The CAT, as with most modems, is compatible with any computer that is equipped with an RS-232 interface. Data exchange can occur at any speed up to 30 characters per second.

Designed for transmitting data over all telephone lines, CAT has many uses for businessmen and hobbyists. Because it allows one computer or terminal to talk to another, a businessman with a CAT modem can work on his payroll, receivables/payables, and inventory right in his home, and hobbyists can talk to each other and even exchange special programs. In short, a CAT modem is the ideal small computer's companion.

For an extra $10.00 though, you can get the D-CAT™ modem. The "D" stands for "direct connect" which means more reliability and even trimmer styling. In fact, this modem is so sleek and trim that some people may not even notice that it is around. It can be installed within a matter of seconds, and it has all the features that the CAT modem has.

Finally, there is even a CCITT CAT that can be used in worldwide computer communications, since it is more compatible with different protocols of modems.

Fig. 6-10. The D-CAT™ modem (Courtesy Novation, Inc.).

Advantages of the CAT and D-CAT Modems
1. Economically priced for the quality of the unit.
2. Has a test mode, an originate mode, and an answer mode.

Disadvantages of the CAT and D-CAT Modems
1. There are no major disadvantages, except for the fact that the modem requires an RS-232 interface.

The AUTO-CAT™ Modem

Another *CAT* modem is the *Auto-CAT*™. It is even more sophisticated, as the $250 price tag indicates. It features automatic answering, low-power operation, and reliable LED indicators which pulsate for tests, power indications, and data transmissions.

The *AUTO-CAT* (Fig. 6-11) has a data rate of 0-300 baud, an EIA RS-232C interface, and is compatible with any Bell 100-series modem, including the *CAT* and *D-CAT* units. Besides communicating data over all telephone networks, *AUTO-CAT* will automatically answer each call, thus allowing unattended operation of personal or business computers. Now, hobbyists can contact their home computers from wherever there is a phone—from across town or across the world. Also, businessmen can communicate with their office computers in the evening, on

---

AUTO-CAT is a trademark of Novation, Inc., Tarzana, CA 91356.

89

Fig. 6-11. The AUTO-CAT™ modem (Courtesy Novation, Inc.).

weekends, or even while on a holiday or vacation. Like the other CAT modems, the AUTO-CAT is so sleek and compact that it fits easily under a telephone with all the controls easily accessible but without being very noticeable.

Not only does the AUTO-CAT feature high baud rates, but also automatic and manual answering, automatic dialing, full or half duplex communication modes, testing modes, a phone line interface, eight front-panel LED indicator lights, and even low (300) baud operation. This is certainly a modem for the more professional computer user, and it is made of the same high-quality parts for which all Novation modems are famous.

Finally, there is the "Cadillac of CATs," the *212 AUTO-CAT*™ modem that is designed for high-baud operation. This modem can transmit data at up to 1200 bits every second, or four times faster than the personal computer standard. Naturally, the price is higher (about $700) but, for business people, time is valuable and this modem can save a lot of it.

Advantages of the AUTO-CATs

1. They have automatic answering, automatic dialing, and high-quality parts.
2. They are low cost, permit unattended operation, are usable on all telephone networks, and are compatible with Bell 100 modems and other CAT modems.
3. They are sleek, trim, and not highly visible.

---

212 AUTO-CAT is a trademark of Novation, Inc., Tarzana, CA 91356.
Touch-Tone is the trademark of American Telephone and Telegraph.
Apple is a registered trademark of Apple Computer Inc., Cupertino, CA, while CAT is a trademark of Novation, Inc., which does *not* manufacture Apple computers.

Disadvantages of the AUTO-CATs
1. The price of these modems, when added to the price of the RS-232 board needed for the *AUTO-CAT*, or to the EIA RS-232 board needed for the *212 AUTO-CAT*, is rather high. However, if you want the best, you will have to pay for it.

APPLE-CAT II™

The *Apple-Cat II*™, according to Novation, Inc., the manufacturer, is "more than just a modem. It's a personal communication system." With it, you can access data banks, swap programs, or talk to your office computer from home. *Apple-Cat II* (Fig. 6-12) organizes your computer memory for memory storage, with incoming messages being held for your convenience. Outgoing messages can also be stored and, then, sent on command.

There are automatic functions that let you set up your Apple to run on its own, including auto dial (either pulse or Touch-Tone™), redial, auto answer, and disconnect. Much of your communications can be done unattended, at night, when line costs are the lowest. You can choose from either standard 110- or 300-baud operation (full duplex), or you can operate at 1200 baud (half duplex), reducing your telephone charges. (Novation also makes what they call an up-grade module, the *212 Apple-*

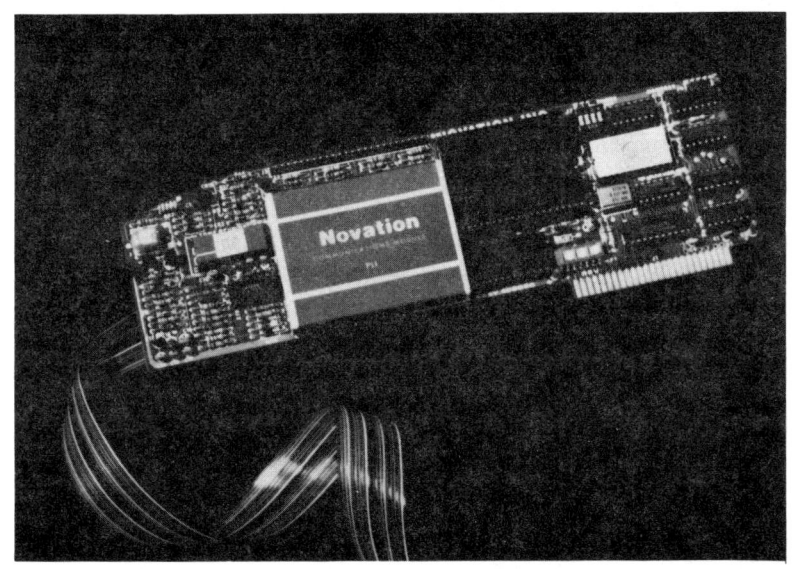

Fig. 6-12. The Apple-Cat II™ printed-circuit board (Courtesy Novation, Inc.).

Cat II^{TM}). By adding the *212 Apple-Cat II* (cost $389) to your *Apple-Cat II* and your Apple computer, you can have full-duplex 1200-baud operation.

The *Apple-Cat II* is also for the deaf community. It has a special 45.5-baud, Baudot-coded, Weitbrecht modem for communicating with the TDD (Telecommunications Device for the Deaf) network.

The serial RS-232 port needed to run your printer is built-in, as is a BSR® X-10 controller. Add a real-time clock to the controller and you will be able to use your computer to program and run home appliances, lights, or anything powered with electricity. Also, even though the *Apple-Cat II* takes up one of your phone lines, you don't have to lose the use of a phone. A standard handset converts your Apple microcomputer into an intelligent telephone at the push of a few keys. It's a handy extra phone to have when you want to precede or follow data transmission with a voice call. You can switch from voice to data any time without losing the connection.

Many of the features of the *Apple-Cat II* are simple add-ons. It is easy to use as all of the functions are fully programmed in. Just insert the supplied diskette, turn the modem on, and your screen displays a menu of options that you can select at the push of a single key. The pc board of the *Apple-Cat II* plugs into one of the peripheral slots inside your Apple (Fig. 6-13). A small interface expansion module goes on the back of the computer and a ribbon cable connects the module to the pc board. A telephone handset and a holder are mounted on the right side of the computer's frame. The plug-in installation is so simple that the entire installation should take no more than 5 to 10 minutes.

Advantages of the Apple-Cat II^{TM}

1. For a basic price of $389 and an amount of about $280 (for all the accessories), you can have what is probably the most advanced modem that you can put into your computer.
2. The modem operates full duplex at 300 baud and up to 1200 baud at half duplex.
3. Adding a *212 Apple-Cat II* up-grade module to your system allows 1200-baud full-duplex operation.

---

212 Apple-Cat II is a trademark of Novation, Inc., Tarzana, CA 91356.
BSR is a registered trademark of the BSR Corporation.
The Weitbrecht modem was developed and designed by Robert H. Weitbrecht, Sc.D., to aid deaf communications.

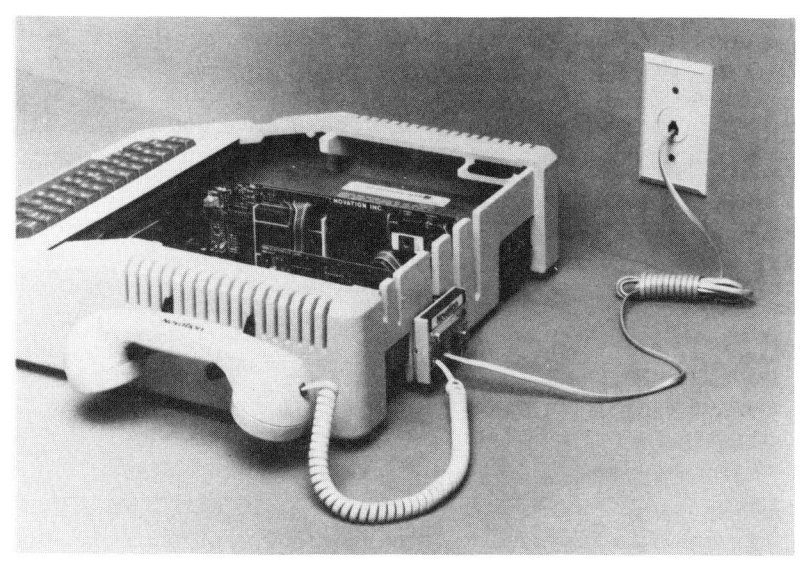

Fig. 6-13. View showing the Apple-Cat II™ installed in the Apple computer, and both the expansion module and the telephone handset/holder mounted on the frame (Courtesy Novation, Inc.).

Disadvantages of the Apple-Cat II™

1. The *Apple-Cat II* requires a 48K Apple II or Apple II Plus with a single disk drive.
2. It also requires a 3.2, 3.2.1, or 3.3 disk operating system (DOS). The accessory disk is formatted in 3.2.1 DOS, but conversion to 3.3 DOS can be done using the Apple II Muffin Program.

Your Buying Decision

You may be thinking that the modems you have just read about are all very much alike. To be honest with you, for the personal computer user, they are. All of them work at 300 baud, and they will all give you the power to communicate with other computers. Still, there are factors that you might want to consider when purchasing a modem. These include:

1. Price. (How much is it worth to you?)
2. Features. (Do you need auto-answer or auto-dial?)
3. Support. (What if you have a problem with your modem? Will the company help you?)

I suggest that you write to the companies that are listed in Appendix C and ask them for information on their modems. Examine carefully the features of each modem (since some new ones may have been added to the ones I have listed here) and think about which modem will be best for you both now and in the future.

The modem is the most important part of your getting into the *world connection,* so the purchase of a modem should be a careful and wise one. If you think before you buy, your modem should open the door to a whole new world for you and should continue serving you for many years to come.

Terminals

If you don't have a computer, but you would still like to be a part of the *world connection,* a terminal may be just the thing that you need. *Terminals* are usually made for the purpose of computer communications and *only* for that purpose. They are less expensive than a computer and a modem, but they offer even more features as far as computer communication is concerned.

There are many, many terminals available, so I can't cover all of them here. However, I will describe what the terminals are generally like, so that you can get a feel for what terminals can do and see if they might be right for you. For your "shopping convenience," I have also included the addresses of some reputable terminal manufacturers in the list of suppliers' addresses in the appendices.

Terminals (Fig. 6-14) are most commonly "dumb." They are called "dumb" because that is what they are—just plain stupid and incapable of performing any of the advanced functions that a modem, such as the Smartmodem, can do. Still, some "dumb" terminals have advantages of their own and some terminals have added features that make them into "smart" terminals.

One of the advantages of terminals is that they are made specifically for data communications, so they are built around that premise only. They have "24 by 80" screens. (These are video displays that have 24 lines, each of which is 80 characters wide.) Along with this, terminals have the ability to transmit from very low baud rates (75 bps) to extremely high, but not always reliable, rates (19,200 bps). In addition, some terminals have advanced features like reverse video, advanced cursor control, and the ability to draw pictures (graphics) on the video screen (Fig. 6-15).

Fig. 6-14. A "dumb" terminal (Courtesy Lear-Siegler, Inc.).

The Infone™

Another distinct advantage is that some terminals are very portable. The "Infone™" by Novation, Inc. (Fig. 6-16) is an example of a very compact terminal that is made for on-the-go business people, investors, and other active people. One version of the *Infone* has been designed specifically as an aid for deaf people. It offers the option of operating at 45.5 baud and Baudot coding, with a Weitbrecht-compatible modem for deaf/TDD communications.

The *Infone* portable terminal is amazingly simple to use, for it can be used to communicate from practically any place that has a telephone line—even a phone booth. It weighs less than 2½ pounds and fits into an ordinary briefcase. There are even extra options which may be added on to it, such as a printer, a cassette recorder, and extra memory. These, too, fit easily into the briefcase. The phone line to the new modular phones is easy to

---

Infone is a trademark of Novation, Inc., Tarzana, CA 91356.

Fig. 6-15. A terminal with advanced graphics (Courtesy Lear-Siegler, Inc.).

unclip and plug into the back of your *Infone* for a direct connection. With older phones and international models, the fold-away acoustic adapter does the job. If you want true convenience, an *Infone*™ is an excellent choice.

The *Infone* shows you how to use the *Infone.* It is easy to use and largely self-teaching so you'll quickly discover that you can operate most functions without going back to the manual. The many things that you can do with an *Infone*™ are too numerous to list here, as are the technical aspects of the machine, but a few are:

1. Use your company computer whether you are at home, in the office, or in the field (maybe clear across the country).
2. Check the status of a customer's orders, his payment records, show him an up-to-date inventory of the items he wants, or check the legal language of a contract with your company's lawyers.
3. Make reservations using auto-dialing and redialing.

Fig. 6-16. The Infone™ (Courtesy Novation, Inc.).

4. Use program and text memory ROMs to store frequently dialed numbers and to remember anniversaries. With the push of a few keys, *Infone*™ will look up a phone number and dial it for you, or flash a reminder whenever a birthday, anniversary, or an appointment time comes.

The *Infone*™ offers other features that seem as though they should be in a science-fiction film. For example, by using the BSR® electrical appliance system, in conjunction with this portable terminal, you could be in California and turn off the lights in your house in New York. That is to say, by using this terminal and this special electrical appliance system, you have control over your entire house, no matter how far away from it you are. (Of course, that's an expensive way to turn off a light but, still, it has possibilities.)

A terminal has a tremendous amount of potential because it is much more portable than a computer, and it has distinct advantages over more expensive computers as far as data communications are concerned. The network of terminals will continue to grow in the *world connection* and, in the years to come, the new technology that is emerging from these advanced devices will be fascinating to see.

Chapter 7
Be a SYSOP

One way to get even more involved in the *world connection* is to become a *system operator,* or *SYSOP,* of a bulletin board service. In this chapter, you will be told how you can become a SYSOP, the equipment you will need, the BBS software you should consider, and the advantages and drawbacks of being a SYSOP.

Presently, there are about 400 SYSOPs in the United States. That number is going to continue to grow, though, and you may want to become a part of it. Being a SYSOP is a great way to meet new people, learn more about your computer, and maybe even make some money.

Getting Started

To run a bulletin board service, you will need several things. First of all, you will need a computer. This is basic enough, but the computer must have several things, also, which includes:

1. At least two disk drives.
2. 48 kilobytes of memory (48K).
3. A printer (not really necessary, but very helpful).
4. An answer-capable modem.

The disk drives are very important, because a BBS needs a lot of space in which to store messages and bulletins, and it must also have speed. There have been cassette-based BBSs, but these were very slow and impractical. The 48K memory is required for the storage of programs that will run the BBS, and the printer, although not absolutely necessary, is good to have around for such things as printing out messages to the SYSOP and "beeping" when someone wants to chat (since the printer has an internal buzzer).

After you have the proper computer equipment, you will

then need to find some good BBS software to run on your system. Most of the BBS software that exists is made for the TRS-80 Model I or Model III microcomputers, while the remainder is primarily for the Apple II microcomputer. However, there are some other systems that are for microcomputers like the PET, the Northstar, and the S-100 computers.

Because BBS software is not extremely popular, software makers have made its price rather high. Good BBS software can range from $100 to $250 in price, but it includes all of the programs and documentation needed to start your bulletin board service. There is an alternative, though, if you happen to be a good programmer; write your *own* BBS software. This not only will save you money, but you might even want to market the software so that you can make some money as well. Keep in mind that such an undertaking requires time and patience on your part.

Choosing a BBS for Yourself

Because there is a relatively small amount of BBSs to choose from, finding a BBS system program for yourself should not be very difficult. The most important thing is to find one that will work with your computer system. After you have done that, select the one with the features that you and your users would most appreciate (Fig. 7-1).

I am going to list some of the popular BBSs along with their features. I will not list most of their prices, however, because they vary from place to place. In fact, most of the time the price for a BBS software program is around $150, so money is not a real factor here because there is little difference in price from one BBS to the next. If you would like the exact price for a particular BBS program, all you have to do is contact the maker of the software by using the addresses I have listed in the appendices.

Connection-80

This BBS, made by BT Enterprises, is for the Radio Shack TRS-80 Model I or III microcomputer, and is one of the most popular BBSs around (both in users and SYSOPs). There are over 100 "Conn-80" systems in the United States, Canada, Mexico, and Europe, so it is certainly an international system.

Fig. 7-1. Select the features you want for your BBS.

One of the main advantages of the *Connection-80* system is its speed. Most bulletin board services take longer to use than the *Connection-80* board; this can be credited to the good programming that went into it. There are commands for bulletins, finding other BBS numbers, ordering products, uploading and downloading, in addition to all of the standard features normally available, such as CHAT and using the "message base" to send and retrieve private messages.

The *Connection-80* is an excellent package, and the program itself is simple to change if you would like to "customize" your BBS. You are certain to get a lot of callers with this system, so I recommend it as one of the best BBSs available.

Forum-80

Bill Abney of Kansas City, Missouri, wrote a program for the TRS-80 several years ago called *Forum-80*. This system is currently being sold by the Small Business Systems Group, Inc. This BBS has been improved and reorganized several times and there are about 50 such systems in the United States. To use this system, you will need a TRS-80 Model I or Model III along with the NEWDOS/80 disk operating system.

When I first called up a *FORUM-80* system, it did not impress me. It was primitive looking, had few extra features, and it was not very well used. However, the latest revision has made it one of the most advanced and popular BBSs in the United States.

There are many extra features to this system. These include advanced uploading and downloading, password protection for messages, and the capacity to store up to 512 users in memory. Another handy feature of the *Forum-80* system is the "obscene filter," which will destroy any messages that have undesirable language in them.

The only real disadvantage to the *Forum-80* software is that it is very expensive for a BBS. The $350 price tag discourages some people from opening up a *Forum-80* system, but if you have the money, it is an excellent system to own and operate.

ABBS

For the Apple user, the *ABBS* is a good choice because it is the most widely used electronic message-system program. You will need only one disk drive for this software, along with your Apple II and a modem.

The *ABBS* does not offer a lot of "frills" as some systems do, but it is an excellent message base. There are a few options available that include uploading and downloading, along with the ability to create "conferences" on the *ABBS*. These are actually messages passed between members of clubs or special interest groups.

Craig Vaughan was the creator of this popular program package. It is being marketed through Software Sorcery, Inc. If you own an Apple computer and you want a very basic BBS, this is the best BBS for you.

Bullet-80

The *Bullet-80* system is not as great in number as other systems, but it is certainly high in quality. The *Bullet-80* board has always been my personal favorite because it has a *lot* of sophisticated extras. In my opinion, this is by far the most advanced board for the TRS-80 Model I or III computer, and many people I have talked with feel the same way.

One of the best features of the *Bullet-80* is that you can change it to your own specifications. You may add a merchandise section (where people can buy things), a magazine section (for articles and items of interest), and anything else you feel would be appropriate for a BBS.

The average *Bullet-80* board gets about 30 callers every day of the year, so it is certainly a good system if you want a lot of callers. It is fast, reliable, and a lot of fun for people to use. I recommend the *Bullet-80* as the best bulletin board for the TRS-80 microcomputer. Give one a call and see what you think.

The People's Message System

This system, almost always called *"PMS,"* was written by Bill Blue for the Apple II Plus computer. You will need an Apple II Plus, a Hayes Micromodem™, and at least two disk drives to run this popular system.

There are many "pluses" in the *PMS* package, including a file for bulletins, sections for advertising and club news, and an "obscene filter." (It will not save any messages with vulgar words in them.) There are even "accounts" which are given to frequent users of the system.

There are not as many *PMSs* as there are other systems, but it is still a good system for the Apple II. Though more expensive than the *ABBS*, the *PMS* offers features that are not found on any other Apple bulletin board service.

Other Boards

There are many other, much less popular, boards which exist today. Many of these are "homemade" boards that rival even the most expensive BBSs. One of these is *"Download-80"* of Concord, California, whose SYSOP is Preston King. This is one of my all-time favorite boards; it was made completely by Preston. Some other good BBSs that are not in the "big-time" yet are the *Mouse-Net* system, the *Living BBS,* and the *Conference Tree.* I suggest that you call some of these to see what they are like, since you may want to be a little different and may want to purchase a board that is not quite as widespread as is the *ABBS* or the *Connection-80.*

If you are the type who likes to read romance novels, perhaps you would like to start a *"Matchmaker"* system. These systems are mainly for the purpose of matching up people—something like computer dating. Every person who calls up the *Matchmaker* must complete a long questionnaire. Then, the computer will match him or her with as many other callers as possible. This is an interesting new twist on the BBS idea. You can see what a *"Matchmaker"* system is by contacting Viva La Difference!, 6065 Roswell Road, NE, Suite 1490, Atlanta, GA 30328. Do not be confused by the names. The original program is called the "Matchmaker," but on various BBSs and in different areas, the program is called many things—"Matchmaker," "Viva La Difference!," or even, "Dial Your Match."

Other Requirements

The only two things that you must remember when buying a system are:

1. Will it work with my computer?
2. Does it have the features I want?

The BBSs I have mentioned here are all quality services, so you will need to be very picky about the features that you will want on your BBS. Happy hunting.

Why Be a SYSOP?

One may wonder why a person would want to invest a total of about $3000 just to have a full-scale BBS, along with all the operational costs and time. True, a BBS is expensive and is time consuming, but there are some distinct advantages to owning and operating a bulletin board service.

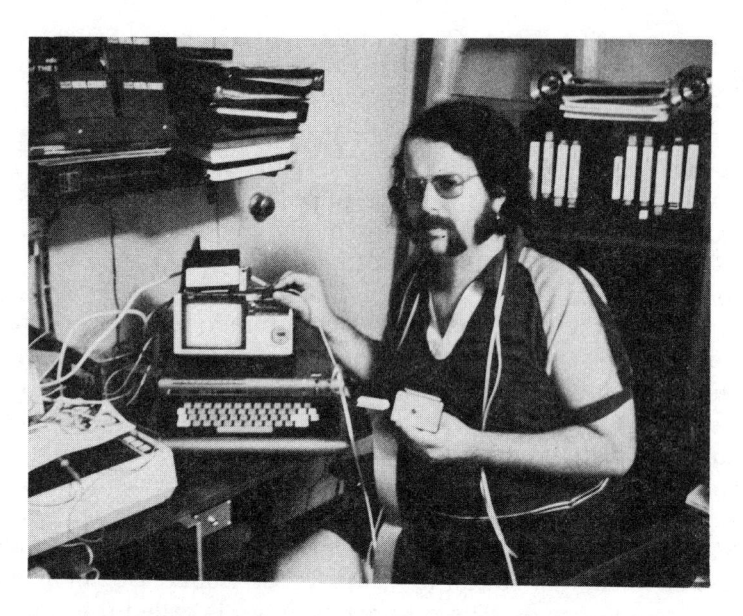

Fig. 7-2. Bob Rosen, the Number 1 SYSOP (Courtesy Bob Rosen).

To show some of these advantages, I am going to tell you about the Number 1 SYSOP in the United States—Bob Rosen (Fig. 7-2). Bob, who lives in Woodhaven, New York, was one of the first people ever to buy a TRS-80, and one of the first to start up a bulletin board service in the New York area. He has gradually built up his system into the most popular one in the country, with over 20,000 calls every year. He has also built up the largest amount of memory space available for a BBS—3 million bytes of on-line disk storage space.

Bob was one of the first TRS-80 owners with the *Connection-80* BBS package, but he soon moved on to even more advanced things, such as the TRS-80 Color Computer. Currently, Bob has two BBSs in his home—one on a TRS-80 Model III and the other on a TRS-80 Color Computer. You can switch back and forth between these two BBSs, and Bob has even more plans for the future of his BBS.

Why does he do all of this? What is the advantage to it? Well, Bob meets a lot of people through his BBS from all parts of the world. He can get advice from them, give advice to them, or just have a friendly chat to see what is going on in their corner of the world (Fig. 7-3).

Fig. 7-3. SYSOPs have all sorts of connections.

Another distinct advantage is that Bob can sell things through his BBS. He has a merchandise section from which people can buy things, like a ROM pack called ColorCom/E (which is needed to download anything from his system) and various other types of software. His mail-order business selling products and advertising through the computer can be quite profitable, so that is another advantage of owning a BBS.

Lastly, the enjoyment derived from owning a bulletin board service is also a plus. Working with electronics, software, hardware, and communicating with a large number of people are all activities that Bob enjoys and which he sees as something positive in running a BBS. As the nation's Number 1 SYSOP, he also knows that bulletin boards are going to continue to improve, and he is going to be there every step of the way in order to enjoy the progress that he and other SYSOPs have made. For more information about Bob Rosen and his bulletin board, see page 174 of the June/July 1982 issue of 80 Microcomputing.

The Drawbacks and Headaches

Being a SYSOP is not an easy task, and there are some definite disadvantages to taking on the job of a SYSOP. Since I have men-

Fig. 7-4. Bulletin Board Services are big business.

tioned that you can make money with a BBS, I would also like to make clear that you also have to spend money on your BBS, for such things as keeping everything running properly, having your system repaired, buying new diskettes, and even paying for the phone bill (if you have a separate phone for your BBS, which is a good idea). On top of that, if you intend to sell things through your BBS, you will have to spend money for owning and operating a business, along with the time you will spend filling out permits and tax forms (Fig. 7-4).

Also, if you are hard pressed for time, a BBS is definitely not for you. Maintaining a BBS takes time; the computer gets a lot more use, so it needs more maintenance. Also, people will want to chat with you every now and then, and you will have to answer the messages that you receive each night. If you enjoy talking to people, these things are all good. However, if you are a busy person, then these are definite drawbacks.

To Be or Not To Be?

Should you be a SYSOP? Ask yourself these questions:

1. Do I have the money to afford the hardware and software?
2. Do I have the time to chat with people and answer messages?
3. Am I willing to sacrifice money and energy to maintain a BBS?

If you can answer "yes" to these three questions, then you are a good candidate for joining the elite few who can call themselves SYSOPs. Think about the prospect and if it entices you, give it a try. I think you will be pleasantly surprised at how much you can get out of it and you will enjoy being a system operator.

Chapter 8
The Soft Side

Just like Chapter 6, this chapter will consist of a number of reviews that are designed to help you select the product which best suits you. The difference here, though, is that this chapter is concerned with software as opposed to hardware. Software, as you may recall, consists of the actual written programs that allow you to communicate with other computers, rather than an electronic "solid" part of the computer system (like a modem).

The software that will be covered here will be for the personal computer (i.e., TRS-80 and Apple II) and is usually priced from $30.00 to $200. The prices do not always reflect the quality of the program, so do not be deceived by them. Some of the less expensive programs are the better ones. Still, keep price in mind so that you can make the choice most cost-effective to you.

Reviewing the Terms

Once again, there are several terms that you should be aware of in order to fully understand these summaries of computer communications software:

Advanced Cursor Control—This term simply means that the cursor (the blinking dot that shows where you are on the video screen) can be moved about the screen easily by means of four arrow keys. These keys (up, down, left, and right) can be used to move the cursor anywhere on the screen, in their respective directions.

Buffer—A buffer is merely a temporary storage space. For instance, if you downloaded a program from a computer system and you were going to save it to a disk, there would have to be a certain area in memory that would serve as your *buffer*. This is because you cannot save files directly from the modem to the disk, so you need a place to put the program temporarily. The term "buffer" can best be summarized as a temporary medium.

Communications Protocols—These are the specific "settings" for your modem, which include parity, bits per word, and duplex. You do not have to understand these terms, but just realize that if you are told to "reconfigure your UART status," you are just being told to use your terminal program to change the settings of your modem. The software will guide you through this process.

Block—This is a block of information. For instance, you might consider each of the chapters in this book to be a "block." In computer communications, though, a block is a certain amount of data, such as 512 bytes. "Block transmission," therefore, is the transmission of data in chunks of data.

All of the other terms that you will need to know were covered in previous chapters so you should now be well-armed to begin your quest for your perfect smart terminal program.

The Software

The vast majority of this software is for, as you might guess, the TRS-80 Model I and Model III microcomputers. I would assume that this is because TRS-80s are used more often in computer communications, but I do not have any statistics to prove that. Still, the market is mainly for the TRS-80, so that is primarily what you will see here.

There is no "perfect" software package for everyone. One terminal program might be perfect for me but would not suit your needs at all and vice versa. We all have different needs so keep that in mind when you are looking at these review summaries. Find out which package suits your needs best.

DDS/DFT

This package consists of two programs, a disk downloading system (DDS) and a direct file transmission (DFT), for the TRS-80 Models I and III microcomputers. They are two completely different programs, but they are made by the same person—Bob Withers of Big Systems Software. The cost for both of the programs is $70.00. There is also a TDS/DFT package for cassette users.

Both of the programs have one main purpose—the transmission of programs. There are no elaborate message buffers or letter editors. These two packages were made for uploading and downloading and that is what they do best.

The DDS has two modes of operation—terminal mode and BASIC mode. You can switch from your computer to the terminal by pressing the CLEAR key. This is a real convenience. Because of this, you can write a BASIC program and, then, transmit it over your modem; all without saving or loading any programs.

Also, you can simultaneously print out on your printer everything that appears on the screen. This is good if you want to record something permanently onto paper. Finally, DDS features both uppercase and lowercase, room for three log-on messages within the program, and user-definable control keys so that you can have the ultimate in convenience while using your TRS-80 terminal.

The DFT program is my favorite part of this package because it can receive and transmit almost any type of program directly, whether it is machine language, word processing files, BASIC, or even assembly language source code. The only problem with DFT is that to transmit programs to someone else, that person must also have the DFT system. However, program transmission is accurate, and each block of data that comes in is simultaneously saved to disk. DFT also has a "chat" mode so that you can talk with the person you contacted after you are finished transmitting data to him.

The TDS/DFT package is one of the lower-priced, but still very powerful, terminal packages. If you need uploading and downloading power, this is the program you are looking for. It has features in transmitting data that few other packages offer, and it is, along with its documentation, well worth its modest price.

The ST-80 Series

The "ST-80" series is a number of programs designed for different versions of the TRS-80 microcomputer. To give you a general idea of the quality of this program, I will describe the package called "ST-80 III." This package is for the TRS-80 Model I or Model III microcomputer, with at least one disk drive.

As you might have guessed, "ST" stands for *smart terminal*, as named by its manufacturer, Small Business Systems Group, Inc. For $150, you can purchase *ST-80 III* along with its extensive 90-page documentation. This system is intended for the "commercial user," since it has top-of-the-line features. These include:

1. Display of free memory left in buffer.
2. Direct cursor control.
3. Serial numbers to discourage piracy.
4. Supports automatic-dialing modems.
5. Can hold large amounts of information that can be printed later onto a printer.

The ST-80 III package also has the standard features of uploading, downloading, and repeating keys. People from NASA, the U.S. Navy, Westinghouse, and many colleges and universities use this terminal program because of its high quality and reliability. If you are going to be using your terminal program for rather serious applications, the ST-80 III package is definitely one that you should investigate.

VisiTerm

This program is for an Apple II computer with at least one disk drive. Made by VisiCorp, the same people who made *VisiCalc*, this program costs $100 and comes with 100 pages of superb documentation and pictures.

A highlight of *VisiTerm* is that it can support many different types of characters. By using the Apple's high-resolution graphics, characters for the APL language (or any language, for that matter) may be created and used on your terminal. This is perfect for world-wide communications, since many languages have different alphabets from ours. The *VisiTerm* program allows your Apple II to communicate with larger computers and other personal computers.

Also, you can change the number of characters within a single line, so you can have a "custom-designed" line length. File transmission is well supported in *VisiTerm,* also, with two main types of file transfer modes—binary mode (which is more compact) and ASCII text mode. You can also adjust your terminal configuration to match the host computer's requirements, such as baud rate, stop bit, parity, and others. Special macros save you time in performing often repeated sequences.

VisiTerm is as high in quality as any of VisiCorp's products, and reviewers have been praising it for some time now. Apple II users should take a serious look into the *VisiTerm* program, since it is one of the highest quality terminal programs for the Apple II.

Smart-80

Yet another terminal program for the TRS-80 microcomputer, this program is made by the makers of the Microconnection modem, and sells for $80.00. It can be provided on either cassette or disk, and there are several versions available for different configurations of computers.

Along with the essentials, *Smart-80* features the uploading and downloading of files from *CompuServe,* the transmission of assembly language files, and two prestored messages to make logging onto a computer system easier.

The *Smart-80* also incorporates a "beep" routine to the keyboard so that typists can make sure that everything they are typing is being entered into the computer properly. Although this is not one of the most popular software packages, it is excellent for owners of the Microconnection modem since it uses the features that are available with the Microconnection to the greatest extent.

OMNITERM

OMNITERM is a program made by David Lindbergh of Lindbergh Systems to work on the TRS-80 Model I and III. A price tag of about $100 includes the price for the program and the instruction manual (which is outstanding). *OMNITERM* is a professional communications package that allows you to easily communicate and transfer files (or programs) with almost any other computer. The package includes the OMNITERM terminal program, four conversion utilities, a text editor, special configuration files, documentation, and user support.

One frustrating thing about computer communications is that text "scrolls" off the screen eventually and it is lost forever. This is not the case with *OMNITERM,* since it can read the information that has scrolled off the top of the screen. Along with this handy feature, *OMNITERM* will reformat the 32-, 40-, and 80-character lines found on many systems to fit neatly on the 64-column TRS-80 video screen. This is especially helpful for electronic mail, since reading an 80-column letter on a 64-column screen can get pretty confusing.

OMNITERM is menu-driven, somewhat like *CompuServe* and *The Source* are, and it is a highly sophisticated program. The manual includes an index, along with a glossary, that makes understanding *OMNITERM* much easier. As full details on OMNITERM cannot be given here, I recommend that you write to Lindbergh Systems (address in Appendix C) to ask for more information about *OMNITERM* and its current price.

UNITERM/80

Apparat, Inc., the maker of the ever popular *NEWDOS/80* disk operating system, has also made *UNITERM/80* for the TRS-80 Model I and Model III. It costs $80.00, and is known by its users as being flexible and unique.

UNITERM/80 supports both uploading and downloading, and a person can examine a downloaded file once the computer finishes transmitting it. The computer can also print the information onto the printer as it comes over the screen. This is done very smoothly. Automatic log-on messages can be used with *UNITERM/80*. This will save you some time when on-line with another computer system.

Since *UNITERM/80* is made primarily for the Microconnection modem, it will first check to see if it is present. If not, it will check to see if another type of modem is connected, such as the LYNX. This is a handy feature, since different modems use different "ports" (or outlets for data). *UNITERM/80* is best for *NEWDOS/80* users, though it will work on most popular disk operating systems.

MODEM 80

MODEM 80, written by Leslie Mikesell of The Alternate Source, is a disk-oriented communications system for the TRS-80 Model I or Model III microcomputer and costs a very low $39.95. This program will permit:

1. Remote operation of a TRS-80 from any terminal via a telephone line.
2. Error-free file transfer with another TRS-80 or even a CP/M computer system.
3. File transfers with large time-sharing systems, such as the CompuServe Information Service and The Source.

This communications package actually consists of seven different programs, each with a function of its own. For example, the program *HEX/CMD* can convert a binary file to hexadecimal (something that is frequently needed in computer communications), and the program *TEXTFIX/CMD* is used for converting text files that are in CP/M format to a TRS-80 format.

The seven programs also come with 35 pages of documentation that thoroughly explain how to use the *MODEM 80*. In addition, it explains some of the terms used in the manual. The program itself is very advanced and has commands for performing DOS commands while using the program and disk access while in terminal mode. It even has commands for sending and receiving graphics characters.

Although the documentation may look skimpy, it covers everything quite well. *MODEM 80* is by far the best value, since it offers the advanced features of many other programs, but at a fraction of the price. Also, The Alternate Source is well known for its excellent customer support and will be willing to answer any of your questions. I highly recommend *MODEM 80* as the best buy for your money.

Super>>Terminal

Instant Software of Peterborough, New Hampshire is the maker of *Super>>Terminal*, a program by David Lindbergh. The program itself sells for $95.00. Accompanied by a very complete 45-page manual, *Super>>Terminal* is probably the best program for electronic mail.

The reason I can say this is because of *Super>>Terminal's* advanced editing features. The program *TEXTED/BAS*, supplied with *Super>>Terminal*, will allow you to create and edit messages before logging onto a system. This can be a real *time and money saver* when on systems like *CompuServe* which cost $5.00 an hour, and even more during the daytime/prime time hours. You can communicate with almost any other computer system quickly and efficiently.

Super>>Terminal also has all of the "standard" terminal features, along with advanced cursor control commands, printer support, and operation at speeds of up to 4800 bits per second (baud). It provides single key auto sign-on, file transfer capabilities, and a continuous count of parity, framing, and overrun errors.

Although Super>>Terminal is more expensive than a program like MODEM 80, it is well-made and well-documented. I recommend this program for those people who intend to use electronic mail often, because it can pay for itself in a very short time with its advanced editing features. However, Super>>Terminal requires a TRS-80 Model I or Model III with 32K memory and at least one disk drive.

Other Programs

Because of the fact that most terminal programs are written for the TRS-80, there are very few smart terminal programs available for computers like the Atari, the Pet, and the Texas Instrument computers, except those available from their own manufacturers. This will certainly change as time goes on, because getting into the *world connection* is becoming more and more popular. The demand for smart terminal software that supports other computers will surely increase.

To find out if there are smart terminal programs available for your particular computer, other than the one(s) that the manufacturer of your computer sells, just go to your local computer store (such as a Computerland) and ask. The people there may just have a program or programs on hand and, if not, can probably tell you where you might get one through mail order.

One type of program, called *VIDEOTEX*, is available for computers like the TRS-80 Color Computer, the Apple, and many others. *VIDEOTEX* is a "skeleton" of smart terminal programs, and its features vary from computer to computer. You might want to inquire to see if *VIDEOTEX* is available for your particular computer. *VIDEOTEX* was designed to access bulletin board services and large commercial data bases. Its graphics capabilities are limited and it cannot access all data bases.

Things To Consider Before You Buy

Buying a smart terminal program can be expensive ($300) or relatively cheap ($30.00). Regardless of price, it is always wise to "look before you leap" into a communications package, so here are some things you might want to consider before you buy a smart terminal program for your computer:

1. Do I need advanced uploading and downloading capability? What kind of programs will I be transferring the most—machine language, word processing files, BASIC?
2. Am I going to be doing a lot of editing with my messages? Do I need advanced cursor control and editing features?
3. What programs will work with my system? If I expand my system, will they still function properly?
4. What can I afford in the line of a communications software package?
5. What types of computers am I going to be communicating with? Will the smart terminal program I am considering work with those computers properly?

Asking yourself these and any other pertinent questions can save you a lot of regret after you purchase your software package.

Be careful when buying software. If possible, try out a software package before you purchase it. Look over the documentation to see if you understand everything and, if not, ask a salesperson for help. Naturally, if you are going to buy through mail-order, you will not be able to do these things. You may save some money that way, but you will not be able to try the program out first.

Buy the software that fits your present and future needs. There are a lot of programs out there—some good, some bad, some overpriced, and some are bargains. Use your good judgement and shopping knowledge to purchase the "soft side" of the *world connection* that will serve you best.

Chapter 9
The World Connected

In this book, I have tried to describe the features that are available to you today in the *world connection*. Many of these things are amazing, and very few people realize all the power that they can have by simply purchasing a terminal or a computer with a modem.

If these electronic wonders of the present day boggle the mind, how much more amazing will the *world connection* of the future be? What possibilities are there in this giant communications revolution? That, precisely, is what we will look at now.

Electronic Mail

Perhaps when you heard about electronic mail, you wondered how this might affect the Postal Service (Fig. 9-1). The Postal Service provides many things for us, but the two main services are the delivery of letters and packages.

First, let's take a look at the package service. Will this be affected by the computer age? At this time, not really. Shipping records and notification of items shipped do change in the process of sending something from one place to another, but since you cannot send material things through a computer, the parcel post service will not be greatly affected at present.

What about the future? "Star Trek" fans may think that something like the "transporter" could be a thing of the future. In case you are not familiar with the transporter, it is a futuristic device that changes matter into energy at the sending station, and, then, changes the energy back into matter again at the receiving station. This is certainly a thing of the far future but, then, who knows? Perhaps, one day, computers may be used to determine the exact structure of an object and, then, be able to reproduce it at another place through the use of

Fig. 9-1. The shape of things to come?

telecommunications. That is certainly something in the far future, though.

As far as the delivery of "regular" mail (such as letters) goes, this could be greatly affected. Right now, a person can use electronic mail to send brief messages to another person with a computer and a terminal. This is very limited, though, so it does not affect the Postal Service greatly.

In the future, however, more advanced electronic mail could certainly put a dent into the U.S. Postal Service's huge industry. Not only will longer letters be possible in the future but, also, there will be letters with pictures (electronically produced), more widespread message delivery, and even the power to save all of your mail onto computer disks, or paper, or both.

Of course, you couldn't send electronic letters to people who do not have access to a computer terminal. However, there are ways around this, such as the Post Office's "mailgram," or electronic mail service. Using this, you can send your letter to the Post Office, where it will be transmitted electronically to its destination, and, then, will be printed and delivered in the normal way. This guarantees overnight delivery, although it is more expensive.

Many people complain about the present-day Postal Service being too slow, too expensive, or too inefficient. I have no major gripes, but I do think that electronic mail is better for many purposes because it is more reliable, often cheaper, and always faster. In fact, electronic mail is almost instantaneous, as compared to the time it takes the Postal Service to deliver a letter.

Fig. 9-2. Communication and learning in action (Courtesy Source Telecomputing Corp.).

There are many stories abounding that the U.S. Postal Service is not going to sit idly by and wait for its business to slowly drop. They, too, are working with data communications and are going to make their system work hand-in-hand with electronic mail. I believe this is a wise move on their part because they are taking a giant like the Postal Service and increasing its power and efficiency by incorporating an emerging technology.

The *world connection* will definitely affect our Postal Service. It may be for better or worse for the Post Office, but it will almost certainly benefit the public. The faster and more reliable mail service that is available through electronic mail will grow more and more popular as a greater number of people purchase terminals and computers. I see electronic mail as a big "plus" of the *world connection,* since it will make communicating with others so much easier (Fig. 9-2).

The Software Industry

Today, there are two primary ways to purchase software. You either go to a computer store and buy it, or you call up a company to have it sent to you through the mail. The buying of software will also change because of computer communications.

One of the ways in which purchasing software has already been changed is through the use of bulletin board services. By that, I mean that you can order a piece of software (or hardware, for that matter) through a bulletin board service. Still, this is not much different than mail ordering so it is no great improvement.

However, in the future, there may be a better way. Perhaps you will be able to call up a computer system, order a piece of software (paying with either a credit card or even a futuristic "debit" card) and, then, have the software downloaded to your system. This means "instant software purchasing" which would be a great convenience to many people. As far as documentation is concerned, that could also be transmitted electronically and simultaneously printed out onto a printer. This type of software buying would be much more efficient, and software prices could be cut drastically because of low overhead (no salespersons or packaging are needed).

Another way that software purchasing could be affected by the *world connection* is through conversations with salespersons. Some day you may be able to call up a computer store and discuss what you are looking for through the use of a modem, rather than having to drive to the store to talk about your software needs. Perhaps you could even see the software actually demonstrated, via your computer terminal, and could decide which package might be best for you.

Many types of consumer purchases might be affected by computer communications, but software buying is almost certainly going to be one of the first.

New Technology

The technology that has made computer communications possible is going to continue to improve over the years and these improvements are going to be amazing in every respect. Developments in terminals, computers, and networks will all take place.

The terminals of today, even the "smart" ones, will all appear to be very dumb to the terminals and computers of the future. Very soon, talking terminals will be available. These terminals will only be able to say a few words at first, such as "READY FOR TRANSMISSION" and "COMPUTER ON" and other short phrases, but, as time goes on, these words will grow to be more complex and more useful. One day, it is likely that all of the data that have been received and stored may be reproduced via a computer that has a voice doing the talking (reproduction).

Voice recognition is also an area that will develop greatly. In case you are not familiar with voice recognition, it is that area of computer technology that is designed to electronically determine what is being said by a person verbally. For instance, if my computer were equipped with voice recognition, I would be able to say "ON" and the computer would turn on. I then might say, "COMPUTER, DIAL NUMBER 827-5549," and it would do just that. This is still a thing of the future, though, since most voice recognition systems used today are rather primitive.

Improvements in message systems (such as BBSs and timesharing systems) will also take place in the years to come. Higher-speed baud operations, such as 1200 baud, will become commonplace. That is four times the speed of present-day communications. This will not only speed up data transmission, but it will also make things like color graphics more practical. Also, multiuser capacity on bulletin board systems will be more frequently seen, since there will be a greater storage capacity for messages and for names on those systems.

Time-sharing systems will offer vastly greater amounts of information and will be able to support thousands of more users. More time-sharing systems will appear, all with greater memory and more efficiency. Also, as more and more people begin to own terminals, special interest groups and users' clubs will continue to grow and increase in popularity.

Technology's Effect

What will these new technologies do to the institutions that we know today, such as newspapers and banking? Data communication is already taking its toll on these two. For example, readers of newspapers have dropped from 69% in 1972 to 57% in 1978. This percentage has continued to fall. It is true that only a small part of this is due to data communications, but as the computer becomes a more practical medium for reading the news, newspapers will continue to lose their readership.

Today, reading the news through a computer is not only impractical, but is also expensive. In years to come, however, news items will be more feasibly retrieved via the computer because of their accessability, timeliness, and lower costs. Cable-tv will also take a bite out of the number of people who read newspapers since tv is widespread enough to allow almost everyone to read the news via a television set.

This will greatly affect the newspaper industry (along with similar industries, such as news magazines). Advertisements could slowly drift from the newspaper to the computer terminal, cutting off 30% to 40% of a paper's advertising income within the near future. Still, newspapers are trying to prevent this from happening to them and are beginning to take an active part in data communications so they won't be left behind in the wake of this revolution.

What about banking? The "banking and investing from home concept is already in practice, though not on a very wide-scale basis. The people who *do use* this type of service seem pleased with it, however. Many users say that ordering products, paying bills, and performing banking transactions from within their own home is a great convenience.

Banking at home is not very widespread now, but what about the future? Expansions and improvements in "banking through your computer" should attract many new customers and could ultimately lead to the point where you can perform almost all of your business through your home computer or terminal. Flight reservations, theatre tickets, and even the buying of groceries, through computer communications, are just a few of the conveniences that are just around the corner. These types of conveniences will hurt some and help others.

Those whom it will help will be the companies that serve as the medium for the communications. The ones that will be hurt are the stores and banks that do not plan ahead and offer conveniences such as computer shopping and computer banking. When the car first became popular, the idea of a "drive-in bank" or a "drive-up food counter" may have seemed laughable. Look around you now and you will see "drive-in" centers everywhere you turn. It is very likely that in the near future you will see the same type of conveniences available using computers rather than cars.

Democracy Threatened

"What about those people who can't afford a computer or a modem? They won't be able to have all of these services and could very well be cheated out of the happier lifestyle that the rest of the public is enjoying."

You may possibly hear someone saying this in the future, since almost anything new that comes along is greeted with some opposition. They may also say that "opinion polls and public reactions may someday be judged solely through computer terminals," so what happens to the feelings of the people who cannot afford that luxury?

True, microcomputers and terminals do cost money. Take a look at the television industry, though. Television sets typically cost about $600. A very inexpensive television costs about $100. Even with these prices, however, there are over 100 million television sets in the United States, or about one television set for every two people. That is about three tv sets in every home. It is safe to assume, therefore, that almost anyone who wants a television set has at least one.

What, then, would prevent almost everyone in the country from buying a computer terminal? You can buy a terminal for less than a tv, so it would seem logical to assume that almost anyone who wanted a computer terminal could get one, especially since it is predicted that prices are going to continue to fall. Therefore, the computer terminal is not a "toy for the rich," but rather a communications medium that can be shared by all. Democracy is safe.

People and Computers

How will microcomputers affect people? Most dramatically, I would think. I know from personal experience that using the computer as a communications medium causes people to treat others differently. This is because:

1. When using a computer, people cannot see each other when "talking." They are no longer self-conscious and, also, they don't have to worry about their appearance.
2. People feel less inhibited about their feelings and expressions such as "kiss" and "hugs" are not in the least bit rare. People seem to be much more open when using a computer to talk with others.
3. Since a large number of people is available to talk to, people can much more readily find others who also share the same interests.

Fig. 9-3. Even marriage ceremonies could change because of the "world connection."

This type of "openness" can lead to some very interesting endings. I have mentioned that I sometimes meet people in person whom I previously knew only through a computer. There have been some much more dramatic happenings, though, such as marriages (Fig. 9-3).

Yes, there have been marriages as a result of people meeting and talking through the computer. This may seem a bit extreme, but some people seem to become such close friends through chatting with each other electronically that they eventually get married. Aunt Nettie of *CompuServe* finds this to be an interesting trend, though she contends that a lady in Los Angeles who married a 60-pound rock is still more interesting.

In the lesser extreme, many friends can be made via computer communications. Knowing people from different parts of the country can also be a rewarding experience, and this type of "branching out" to others could make the country on the whole a lot closer together in spirit.

Relationships are going to change because of the *world connection*, there is no doubt. I believe that the change will be for the better, which is yet another advantage of this new and exciting revolution.

Fig. 9-4. The world awaits you.

Final Thoughts on Things To Come

The *world connection* is going to change our lives. Life will become more convenient, though more fast-paced, and you will meet people from all over the world with the same interests as you. You will find out things about yourself and your computer that you never knew before. You might find new hobbies and interests that you can develop because of this new communications medium (Fig. 9-4).

There will be many other positive changes because of this revolution. People should become more open with each other and more aware of their own feelings. I know that my own life has become more enjoyable and productive because of my connection to the outside world through my computer and my modem.

Use your computer as an extension of something far more powerful—your mind. The modem can be your magic carpet to a world that you have never known before. Come explore the *world connection* . . .; it's one revolution that you are bound to use and enjoy.

Appendix A
Glossary of Terms

ABBS—The Apple Bulletin Board Service, one of the most popular BBSs in the United States. The ABBS was written by Bill Blue and Craig Vaughan for the Apple II microcomputer.

Acoustic—This is a type of modem. Also, pertains to sound. An "acoustic modem" is the type where you must press the handset of the phone into two rubber cups. The modem then sends and receives data directly through the telephone handset. Outside noise is sometimes a problem with these modems as it can affect the data being transmitted.

ASCII—An acronym for American Standard Code for Information Interchange. This is the method used for encoding digital signals and is usually one of the few things that most personal computers have in common with each other.

Baud—Has the same meaning as bits per second. This is the rate or measure of transmission speed in data communications. For instance, 300 baud means that every second 300 bits of information are transmitted.

BBS—Abbreviation for a computer Bulletin Board Service, usually operated on a personal computer and owned by a private individual.

Bit—The smallest unit of information that a computer can store. A bit is usually designated as either "1" or "0," and a series of eight bits makes up a "byte," which equals about one letter or number.

Block—A chunk of information. A "block" of a program is something like a chapter of a book; it is a part of the whole.

Bps—Abbreviation for bits per second. See *baud*.

Buffer—A place used for the temporary storage of programs, messages, or anything else. The buffer is usually with the RAM (random-access memory).

Byte—A group of eight bits. A byte usually makes up the equivalent of a letter, a number, a character, or some other single piece of information.

Common Carrier—Companies that serve the public and can be used for the transmission of data (such as the telephone companies).

CompuServe—Currently the largest personal computer information network in the United States. *CompuServe* serves thousands of people and can provide information on almost any subject.

CPU—Abbreviation for central processing unit; the "brain" of a computer.

Cursor—Usually a flashing block that shows where the next character will appear. This helps you to know your "position" on the screen.

Data—Any information, such as programs, messages, or numbers.

Direct Connect—A method of connecting something; i.e., a modem that is connected directly to the telephone line. This type of modem is slightly more expensive, but it is more accurate in data transmissions. Compare with *acoustic*.

FORUM-80—Bill Abney's very popular bulletin board service.

Full Duplex—Technically, the ability to "talk" both ways simultaneously. Practically, full duplex is the condition in which you can see the things that you type on the screen of the monitor while on-line with another computer.

Half Duplex—A state in which you cannot see the things you type onto the screen of the monitor while on-line with another computer. A connection that requires end terminals to take turns in passing data—first one and, then, the other.

Hardware—The actual mechanical and electronic part of a computer. A modem is a piece of hardware. See *software*.

K—An abbreviation for "kilo." When used with microcomputers, it stands for 1024 bytes. For example, if you have a 48K computer, you have a computer that has over 48,000 bytes of memory you can use.

Modem—Abbreviation for modulator/demodulator. The piece of hardware that allows a computer to enter "the world connection" by translating electronic signals into audio tones and, then, sending the tones over a telephone line to another computer for "demodulation" by that computer.

Network—A computer system that can support many terminals at once. For instance, the CompuServe Information Service is a network because thousands of people can access the main computer at the same time.

On-line—The state of being connected with a computer network or a BBS.

PMS—Abbreviation for the People's Message System, a BBS by Bill Blue that runs on the Apple II microcomputer.

RAM—The "random-access memory" of a computer. This is the memory that may be used by the user for storage of programs and data, as opposed to a ROM.

ROM—The memory for a computer that cannot be used by a person directly, and which cannot be changed.

RS-232C—An interface device for computer communications, which is considered as the standard and which is usually used to hook a microcomputer up to a modem. Sometimes the RS-232C interface is included with a microcomputer, but often you must purchase it separately for around $100.

Software—Programs stored on some type of media, usually disks or tapes. Software, as opposed to hardware, is the nonelectronic part of the computer. It tells the computer what to do. For instance, a communications *program* is a piece of *software*. See *hardware*.

THE SOURCE—An information network owned by Reader's Digest, Inc. THE SOURCE provides information to thousands of users. It is one of the largest networks presently being used in the nation.

ST-80—A communication program series written by Lance Miklus.

SYSOP—Abbreviation for system operator. It refers to the owner and operator of a bulletin board service.

Terminal—A piece of hardware designed exclusively for computer communications, or a computer being used to communicate with other computers.

Appendix B
Bulletin Board Service Numbers

As of September, 1982, there were about 400 bulletin board services across the country. I have listed about 100 of those services here, so that you can call up one in your area and, perhaps, get the list of all 400 from that one system.

Most of these bulletin board services will be in operation for some time, but numbers are constantly changing. Some boards go down, others change telephone numbers, and others just seem to vanish. If you encounter a number that is no longer a BBS, just cross it off your list and try another one in your area. Enjoy!

BBSs

Some of the computer bulletin board systems of the United States:

Name	Location	Telephone Number
ABBS	AKRON, OH	216-745-7855
ABBS	ATLANTA, GA	404-953-0723
ABBS	BUENA PARK, CA	714-739-0711
ABBS	CAMBRIDGE, MA	617-354-4682
ABBS	CHICAGO, IL	312-622-9609
ABBS	CHICAGO, IL	312-337-6631
ABBS	DALLAS, TX	214-634-2668
ABBS	DENVER, CO	303-759-2625
ABBS	DOWNERS GROVE, IL	312-964-7768
ABBS	FT LAUDERDALE, FL	305-566-0805
ABBS	FT WALTON BEACH, FL	904-243-1257
ABBS	HAWTHORNE, CA	213-675-8803
ABBS	HOUSTON, TX	713-977-7019
ABBS	IRVINE, CA	714-751-1422
ABBS	LOS ANGELES, CA	213-349-5728
ABBS	MARINA DEL REY, CA	213-821-7369
ABBS	MEMPHIS, TN	901-761-4743
ABBS	MIAMI, FL	305-821-7401
ABBS	MINNEAPOLIS, MN	612-929-8966
ABBS	NAPERVILLE, IL	312-420-7995
ABBS	PHOENIX, AZ	602-957-4428
ABBS	POMPTON PLAINS, NJ	201-835-7228
ABBS	SAN DIEGO, CA	714-582-9557
ABBS	SAN FERNANDO, CA	213-340-0125
ABBS	SAN FRANCISCO, CA	415-948-1474
ABBS	SANTA MONICA, CA	213-394-1505
ABBS	SEATTLE, WA	206-524-0203
ABBS	SPRINGFIELD, MO	417-862-7852
ABBS	WEST PALM BEACH, FL	305-689-3234
ABBS	WESTMINSTER, CA	714-898-1984

Name	Location	Telephone Number
CBBS	ATLANTA, GA	404-394-4220
CBBS	CHICAGO, IL (H.Q.)	312-545-8086
CBBS	DETROIT, MI	313-288-0335
CBBS	LOS ANGELES, CA	213-843-5390
CBBS	PORTLAND, OR	503-646-5510
CBBS	WASHINGTON, DC	703-281-2125

Other Important Systems

Name	Location	Telephone Number
DOWNLOAD-80	CONCORD, CA (The SYSOP is Preston King)	415-827-5549
MSG-80	AKRON, OH	216-724-1963
MSG-80	HALEDON NJ	201-790-6795
MSG-80	LIVINGSTON, NJ	201-992-4847
MSG-80	MANHATTAN, NY	212-245-4363
PMS	ANAHEIM, CA	714-772-8868
PMS	FREEPORT, TX	713-233-7943
PMS	LOS ANGELES, CA	213-291-9314
PMS	PALO ALTO, CA	415-493-7961
PMS	SAN DIEGO, CA	714-582-9557
PMS	SANTEE, CA (H.Q.)	714-443-8754
NORTH*	ATLANTA, GA	404-939-1520
NORTH*	COLUMBIA, SC	803-771-0922
BUL-80	DANBURY, CT	203-744-4644
BUL-80	SPRINGFIELD, MO	417-529-1113
CON-80	FREMONT, CA	415-651-4147

The FORUM-80 Network Systems

Name	Location	Comments	Telephone Number
FORUM	ALBANY, NY	(3.1.3)	518-785-8478
FORUM	AUGUSTA, GA	(3.1)	803-279-5392
FORUM	BOSTON, MA	(3.1)	617-431-1699
FORUM	CHARLESTON, SC	(3.1)	803-552-1612
FORUM	CLEVELAND, OH	(3.1)+	216-486-4176
FORUM	DENVER, CO SYS #1	(3.1)	303-341-0636
FORUM	DENVER, CO SYS #2	(3.1)	303-399-8858
FORUM	FAIRFAX, VA	(3.1)	703-978-7561
FORUM	FT LAUDERDALE, FL	(3.1)	305-772-4444
FORUM	HULL, ENGLAND	(3.1)	011-44-482-859169
FORUM	KANSAS CITY, MO	(3.1.3)+	816-861-7040
FORUM	KANSAS CITY, MO	(3.1.3)+	816-931-9316
FORUM	LAS VEGAS, NV	(3.1)	702-362-3609
FORUM	LEAVENWORTH, KS	(3.1)	913-651-3744
FORUM	MEDFORD, OR	(3.1)	503-535-6883
FORUM	MEMPHIS, TN	(3.1)	901-276-8196
FORUM	MEMPHIS, TN	(3.1)+	901-362-2222
FORUM	MONMOUTH CTY, NJ	(3.1)	201-528-6623
FORUM	MONTGOMERY, AL	(3.1)	205-272-5069
FORUM	MT CLEMENS, MI	(3.1)	313-465-9531
FORUM	NASHUA, NH	(3.1)	603-882-5041
FORUM	ORANGE COUNTY, CA	(3.1)	714-952-2110
FORUM	PONTIAC, MI	(3.1)	313-335-8456
FORUM	PRINCE WM CTY, VA	(3.1)	703-670-5881
FORUM	SAN ANTONIO, TX	(3.1)	512-340-6720
FORUM	SEATTLE, WA	(3.1)	206-723-3282
FORUM	SHREVEPORT, LA	(3.1)	318-631-7107
FORUM	TAMPA, FL	(3.1)	813-935-8428
FORUM	TULSA, OK	(3.0)	918-747-1310
FORUM	UNION, NJ	(3.1)+	201-688-7117
FORUM	WESTFORD, MA	(3.1)	617-692-3973
FORUM	WICHITA, KS	(3.1)	316-682-2113
FORUM	WICHITA FALLS, TX	(3.1)	817-855-3916

Meanings of Names and Abbreviations

The names:

ABBS is the Apple Bulletin Board Service.
BUL-80 is the Bullet-80 system.
CBBS is the Computer Bulletin Board Service.
CON-80 is the Connection-80 system.
FORUM is the FORUM-80 system.
NORTH* is the Northstar System.
MSG-80 is the Message-80 BBS.
PMS is the People's Message System.

The abbreviations:

"H.Q." means the headquarters of the BBS system.
"3.1" means the Version 3.1 of FORUM-80 (recent).
"+" means that it will support a 1200-baud rate.

Appendix C
Addresses Of Hardware/Software Suppliers

This list of suppliers is given in order to help you find different services and/or materials. It should not be considered as an endorsement of any supplier; you should choose the one(s) you prefer. The suppliers that *do not* have a telephone number listed with their address would prefer that you contact them by mail.

The Alternate Source (Modem 80)
704 N. Pennsylvania Ave.
Lansing, MI 48906

Apparat, Inc. (UNITERM/80, TRS-80 hard-
4401 S. Tamarac Parkway ware and software)
Denver, CO 80237
(800) 525-7674

Big Systems Software (DDS/DFT software)
P.O. Box 405
Fraser, MI 48026

Bill Blue (People's Message System
P.O. Box 1318 software)
Lakeside, CA 92040

BT Enterprises (Connection-80 software)
171 Hawkins Road
Centereach, NY 11720

CompuServe Information Service (CompuServe)
Personal Computing Division
5000 Arlington Centre Blvd.
Columbus, OH 43220
(614) 457-8600

Emtrol Systems, Inc. (LYNX modem)
123 Locust Street
Lancaster, PA 17602
(717) 291-1116

Hazeltine Corporation (Terminals)
10 East 53rd Street
New York, NY 10022

Hayes Microcomputer Products, Inc. (Modems)
5835 Peachtree Corners East
Norcross, GA 30092

Instant Software, Inc. (Super Terminal)
Peterborough, NH 03458

Lance Miklus, Inc. (ST-80 software)
217 South Union Street
Burlington, VT 05401

Lear Siegler, Inc. (Terminals)
714 North Brookhurst Street
Anaheim, CA 92803

Lindbergh Systems (OMNITERM software)
41 Fairhill Road
Holden, MA 01520
(617) 852-0233

MCI Telecommunications Corp. (Alternate phone service)
1150 17th Street, NW
Washington, DC 20036
(202) 872-1600

The Microperipheral Corporation (The Microconnection)
P.O. Box 529
Mercer Island, WA 98040
(206) 454-3303

MSI Data Corporation (Portable Terminals)
340 Fischer Avenue
Costa Mesa, CA 92626
(714) 549-6000

NOVATION, Inc. ("CAT" modems)
18664 Oxnard Street
Tarzana, CA 91356

PLATO (College information
Contact a local college or university system)
for information regarding this utility.

Radio Shack (Modems and Software)
A Division of Tandy Corp. See your local Radio
Fort Worth, TX 76102 Shack store.

Small Business Systems Group, Inc. (ST-80 and FORUM-80
6 Carlisle Road software)
Westford, MA 01886
(617) 692-3800

Software Sorcery, Inc. (ABBS software)
7927 Jones Branch Drive, Suite 400
McLean, VA 22102
(703) 385-2944

Southern Pacific Communications Co. (SPRINT telephone
P.O. Box 974 service)
Burlingame, CA 94010
(415) 692-5600

TYMNET, Inc. (Common carrier)
20665 Valley Green Drive
Cupertino, CA 95014
(408) 446-7000

Viva La Difference! (Computer dating
6065 Roswell Rd., NE system)
Suite 1490
Atlanta, GA 30328

Index

A

ABBS, 63, 101, 103, 127, 132
Abney, Bill, 101, 129
Access codes, illegal, 72
Acoustic modem, 22, 75-76, 84-85, 127
Advanced cursor control, 108
Advantages of the BBS, 61-63
Adventure, 37, 44
Alternate
  phone service, 72
  Source; see The Alternate Source
  telephone service, 27
American Telephone and Telegraph, 28
Apparat, Inc., 113
Apple-Cat II, 91-93
ARTGAL, 47
Aunt Nellie, 43, 48, 52-53, 124
Auto-
  answer modem, 75, 84
  CAT modem, 89-91
  dial modem, 75, 84

B

Banking, 122
Barter Worldwide, Inc., 37
Bartering, 37
Baud, 23, 128; *see also* bps
BBS, 13-14, 16, 56, 64, 69, 70, 98, 103, 121, 128; *see also* bulletin board service
  advantages of, 61-63
  calling the, 57-58
  menu of a, 13
  software, 98-99
  system program, 99
  using wisely, 63-64
  what is a?, 56-57
Bell
  103-type modems, 82, 83
  212A-type modems, 82
  System, 21, 25-26
Big Systems Software, 109
Binary digit, 23

Bit, 20, 23, 128
Blackjack, 37, 44
Bleeder boxes, 72
Block transmission, 109
Blue, Bill, 102, 127, 130
Bps, 23, 128; *see also* baud
BSR
  Corporation, 92
  electrical appliance system, 97
  X-10 controller, 92
BT Enterprises, 99
Buffer, 108-109, 128
Bullet-80, 63, 102, 133
Bulletin(s), 58
  board services, 16, 26, 56, 57-58, 63, 69, 70, 98, 115, 120
  future of, 64
Business
  and educational journals, 40
  and financial
    markets, 35-36
    services, 43, 46
  computing, 34
Buying services, 37, 44
Byte, 23, 128

C

Calling the BBS, 57-58
Career and education, 35, 37-38
Carrier signal, 25
Case switch, 58
CAT modem, 87-89
Catalog shopping, 35, 36-37
CB
  commands, 51-52
  simulation, 44
Citizen's Band radio, 43, 48-52
Chat, 60, 61, 63
  mode, 38, 110
Classified ads, 37, 38, 43, 44
Commands, 58-61
Common carrier, 25, 128
Communications
  addiction, 19-20
  and mail, 35, 38

139

Communications—cont
 computer, 11-18, 26, 38, 59, 65, 69, 70, 88, 108, 112, 120
 data, 20, 21, 94, 121, 122
 package, 114
 protocol, 109
 satellites, 21
CompuServe, 47-48, 49, 112-114, 124, 128
 Information Service, 13, 29, 30, 41-54, 55
 subject index, 43-44
Computer(s)
 and people, 123-124
 communications, 11-18, 26, 38, 59, 65, 69, 70, 88, 108, 112, 120
 crime, 65-66
 newsletters, 46
 systems, 15
 time-share, 13, 29, 30
Computing and creating, 35, 39-40
Compu-U-Star, 37, 44, 64
Conference Tree, 103
Connection-80, 99-100, 103
Copyright infringement, 68
Creating and computing, 35, 39-40

D

Data
 banks, 28
 bases, 26
 communications, 20, 21, 28, 94, 121, 122
 mode, 25
 transmission, 22, 26
D-CAT modem, 88-89
DDS/DFT, 109-110
Decwars, 53
Direct-connect
 modem, 22, 75-76, 84
 Modems I and II, 85-87
Documentation, 71
Dow Jones Information Retrieval Service, 54
Download-80, 103, 133
Downloading, 58-59, 62, 64, 69
Duplex operation, 23-24

E

80 Microcomputing, 105
Education and career, 35, 37-38
Educational journals, 40
Electronic
 mail, 18, 20, 29, 34, 38, 44, 49, 114, 117-119
 Yellow Pages, 28
EMTERM terminal program, 78-79
Emtrol Systems, Inc., 76
Expert user, 60

F

Financial
 and business services, 43, 46
 markets, 35-36
Foreign language instructional programs, 37-38
Forum-80, 63, 101, 129, 134
Full
 and half duplex switching, 80
 duplex, 23-24, 129

G

Games, 37, 44
Goodbye, 59

H

Half duplex, 23-24, 129
Hardware, 15, 16, 40, 47, 129
Hayes
 Microcoupler, 82, 83
 Micromodem II, 82-84
 Smartmodem, 79-81
 Smartmodem 1200, 81-82
 Stack, 79
 Terminal Program, 84
Help, 59
Home
 and leisure, 35, 37
 computer crime, 66
 services, 43, 44-46
Host computer, 57, 60

I

Identification number, 33-34
Illegal access codes, 72

Infone, 95-97
Information utility, 31, 34
Instant Software, Inc., 114

K

Kill message, 59
King, Preston, 103, 133

L

Leisure and home, 35, 37
Lindbergh,
   David, 112, 114
   Systems, 112, 113
Living BBS, 103
Long-distance phone switching
   system, 21
LYNX, 76-79, 113

M

Mail
   and communications, 35, 38
   electronic, 18, 20, 29, 34, 38,
      44, 49, 114, 117-119
Mailgram service, 41', 118
Matchmaker system, 103
MCI, 27, 72
Menu(s), 34, 43
   -driven, 113
   of a BBS, 13
Micklus, Lance, 79, 130
Mikesell, Leslie, 113
Modem(s), 11-13, 15, 20, 21, 23, 56,
      69, 71, 73, 93-94, 126, 129
   I, 85-86
   II, 86-87
   80, 113-114
   212 Apple-Cat II, 92
   212 AUTO-CAT, 90
   a multitude of, 75-97
   acoustic, 22, 75-76, 84-85
   Apple-Cat II, 91-93
   AUTO-CAT, 89-91
   CAT, 87-89
   D-CAT, 88-89
   direct-connect, 22, 75-76, 77, 84
   from Novation, Inc., 87-93

Modem(s)—cont
   Microconnection, 112, 113
   Weitbrecht, 92, 95
   modulator/demodulator, 21, 129
   Mouse-Net system, 103
   Multi-Player Host, 43, 48, 52, 53
   Multiuser capability, 64

N

National bulletin board, 44
Network systems, 13, 30
News
   and reference resources, 35
   and Reports, 46
Newsletters, computers, 46
Newspapers, 43-44
Novation, Inc., 95-97
   modems from, 87-93

O

OMNITERM, 112-113
On-line transmission, 24

P

Password, 33-34, 73
People and computers, 123-124
People's Message System, 102, 130,
      133
Personal computing, 34
   services, 43, 46-47
Phone
   phreaking, 66, 71-73
   switching system, 21
Piracy, software, 66-71
Pirate
   boards, 69-70
   Fests, 70
PLATO National Educational
   Computer System, 55
PMS; see People's Message System
Popular Electronics, 46
Programs
   protected, 70-71
   smart terminal, 15
   trading, 68, 70
Programming languages, 46-47
Protected programs, 70-71

Q
QUBE, 28

R
Radio Shack, 84, 85, 87
Reader's Digest Association, 31, 130
Reference
   Databases, 46
   library, 44
   resources, 35
Research section, 44
Rosen, Bob, 104-105

S
Satellites, communications, 21
Serial, 10
Software, 15-16, 40, 47, 69, 108-116, 130
   cracking, 70-71
   industry, 119-120
   piracy, 66-71
   Sorcery, Inc., 101
Sound, 22
Source; see The Source
   Plus, 35, 40
   Telecomputing Corp., 31
Sourceworld, 53
Southern Pacific Communications Company, 27
Space War, 53
SPRINT, 27, 72
Small Business Systems Group, Inc., 101, 111
Smart
   -80, 112
   terminal programs, 15, 116
ST-80 Series, 110-111, 130
Subject index, 43-44
Super >> Terminal, 114-115
SYSOP, 64, 69, 98-107, 130
   be a, 16
   drawbacks and headaches of, 105-106
   talk with, 60, 63
   why be?, 103-105
System operator, 98; *see also* SYSOP

T
Tandy Corporation, 84
Technology
   effect, 121-122
   new, 120-121
Telecommunications Device for the Deaf, 92, 95
Telecomputing Corporation of America, 31
Telephone(s), 20-21, 22
   costs, 26-27
   Interface II, 84-85
   network, 73
   service, 28
      alternate, 27
   systems, 20-21
Terminal(s), 94, 97, 116, 120, 130
The Alternate Source, 113, 114
The Source, 13, 14, 29, 30, 31-41, 43, 53, 55, 113, 114, 130
Time-share
   computer systems, 13, 29, 30
   systems, 69, 121
Today, 53
Trading programs, 68, 70
Travel and dining information, 37

U
United Press International news service, 35
UNITERM/80, 113
Upload, 60, 62

V
Vaughan, Craig, 101, 127
Video displays, 23, 94
Videotex, 27-28, 115
VisiCorp, 111, 112
VisiTerm, 111-112
Viva La Difference!, 103

W
Weitbrecht
   Robert H., Sc.D., 92
   modem, 92, 95
What is a BBS?, 56-57
Withers, Bob, 109

TO THE READER

Sams Computer books cover Fundamentals — Programming — Interfacing — Technology written to meet the needs of computer engineers, professionals, scientists, technicians, students, educators, business owners, personal computerists and home hobbyists.

Our Tradition is to meet your needs and in so doing we invite you to tell us what your needs and interests are by completing the following:

1. I need books on the following topics:

_____
_____
_____
_____
_____
_____
_____

2. I have the following Sams titles:

_____
_____
_____
_____
_____
_____

3. My occupation is:

_____ Scientist, Engineer	_____ D P Professional
_____ Personal computerist	_____ Business owner
_____ Technician, Serviceman	_____ Computer store owner
_____ Educator	_____ Home hobbyist
_____ Student	Other _____

Name (print) _____

Address _____

City _____ State _____ Zip _____

Mail to: **Howard W. Sams & Co., Inc.**
Marketing Dept. #CBS1/80
4300 W. 62nd St., P.O. Box 7092
Indianapolis, Indiana 46206

22042